DIY SUSTAINABLE HOME PROJECTS

80+ IDEAS FOR SUSTAINABLE LIVING

SAM FURY

Illustrated by
NEIL GERMIO

Copyright SF Nonfiction Books © 2020

www.SFNonfictionBooks.com

All Rights Reserved
No part of this document may be reproduced without written consent from the author.

WARNINGS AND DISCLAIMERS

The information in this publication is made public for reference only.

Neither the author, publisher, nor anyone else involved in the production of this publication is responsible for how the reader uses the information or the result of his/her actions.

CONTENTS

Introduction   xi

FOOD

VEGETABLE GARDENING

Efficient Gardening	5
Composting	10
Compost Tea	13
Drip Irrigation	14
Harvesting and Storing Seeds	16
Regrow Vegetables from Scraps	18
Natural Insecticides & Herbicides	19
Greenhouses	22
Garden Beds	24
Raised-Bed Garden	26
Vertical Garden	28
Three Sisters Garden Plan	31
Permanent Mulch Gardening	33
Keyhole Garden	35
Hydroponics	37
Aquaponics	41

SOY

Oil	47
Soy Flour	50
Soy Milk	51
Soy Yoghurt	53
Tofu	54
Miso	58
Soy Sauce	61
Doenjang and Guk-Ganjang	63

CHICKENS

Raising Chickens	69
Chickens for Eggs	73
Automatic Watering Tube	78
Chicken Tunnel	80
Breeding Chickens	82
Styrofoam Incubator	86
Chicken Bathing	90
Culling Chickens	91

BEE KEEPING

Starting Your Bee Hive	101
Collecting Honey and Wax	104

PRESERVING FOOD

Drying	109
Drying Fruit	112
Fruit Leather	114
Drying Vegetables	115
Drying Herbs	116
Powdered Eggs	118
Potato Flakes	120
Cardboard Box Solar Dehydrator	122
Pickling	124
Vinegar	126
Apple Cider Vinegar	128
Pickled Eggs	130
Fermenting	131
Sauerkraut	132
Kim Chee	134
Canning	136
Freeze Drying	140
Root Cellar	142

COOKING

Solar Oven	147
Thermal Cooker	151
Portable Rocket Stove	152
Mud Oven	154

WATER

Rain Barrel	165
Shallow Well	168
Fog/Mist Harvesting	175
Water Testing	177
Gravity Ceramic Filter	179
SODIS	181
Bio Sand Filter	183

FUELS

Beeswax Candle	191
Firewood	193
Charcoal	196
Bio-Bricks	198
Gel Fuel	200
Biogas	202

ALTERNATE ENERGY

Alternative Energy	211
Soda Bottle Light Bulb	215
Pedal Power	217
Pedal Generator	219
Solar Powered Appliances	222
Simple Solar Setup for DC Appliances	225

HEATING AND COOLING

Portable Soda-Can Solar Heater	229
Solar Hot Water Shower	232

HEALTH AND HYGIENE

Castile Soap	239
Lye	242
Liquid Castile Soap	245
Cold Pressed Castile Soap	248
Deodorant	250

Sunscreen	251
Lip Balm	253
Cayenne Salve	254
Herbal Medicine	256
Fire Cider Tincture	258
Pest Control	260
Cockroach and Ant Killer	261
Mosquito Ovitrap	263
Rodent Bucket Trap	265
References	269
Author Recommendations	271
About Sam Fury	273

THANKS FOR YOUR PURCHASE

Did you know you can get FREE chapters of any SF Nonfiction Book you want?

https://offers.SFNonfictionBooks.com/Free-Chapters

You will also be among the first to know of FREE review copies, discount offers, bonus content, and more.

Go to:

https://offers.SFNonfictionBooks.com/Free-Chapters

Thanks again for your support.

INTRODUCTION

Whether you live rurally, in the suburbs, or in the city, you will find projects in this book for creating a more sustainable home.

Most of the projects are relatively small. The idea is that you build the small version first to see if it will work for you. Then. if you want, you can expand on it. Many of the projects are directly scalable. For the more technical ones, you may need to do some deeper research or hire some help.

Why Have a Sustainable Home?

There are many reasons why you should make your home as sustainable as possible. Here are a few common ones:

- **Save money.** When you provide something for yourself, that's money you don't have to spend to buy it. It won't be completely free, but you will be bypassing all the middlemen, and that can mean huge savings.
- **Be healthier.** There are many preservatives and chemicals in most commercial food and sundries. If you grow and make your own, you can leave all the harmful stuff out.
- **Help the environment.** Maintaining a sustainable home is a minimal-waste lifestyle, and the less waste you produce, the better it is for the environment.
- **Prep.** If a disaster situation lasts long enough, there will be a shortage of necessities. The more sustainable your lifestyle is, the less you will need to worry about it. You may even be producing enough so you can help others in need.
- **Create extra income.** There are many ways you can make extra income from your sustainability projects, such as selling extra produce, making and selling natural sundries, holding sustainable living courses, and whatever else you come up with.

- **Have fun.** Whether or not you find these sustainable projects fun or not depends on you. At the very least, they will keep you busy.

Using this Book

This book is split into six sections. Each one has an introduction with general information. After the introduction are the DIY projects related to that section.

- **Food.** How to create, cook, and preserve your own food.
- **Water.** Water collection and treatment.
- **Fuels**. Different types of fuels you can use to create heat, gas, fire, etc.
- **Alternative Energy**. Creating alternative energy systems to power your home.
- **Heating and Cooling**. Ways to heat and cool your home without electricity.
- **Health and Hygiene**. Making your own health- and hygiene-related products, including natural pest controls.

Some sections have several projects that achieve the same goal (multiple methods of growing food, for example). Compare them to decide which is the best for you.

Once you have selected a project, read the instructions in full before starting. Get a good understanding of the process and plan for any modifications you want to make (using replacement materials or making things in different sizes, for instance). Do further research if you need to.

Most of the projects follow a similar format:

- **Introduction.** This will give you an overview of the project and how it works. It will also include any relevant safety concerns.
- **What you need.** A list of the materials and tools you need

to make the project. You can substitute or modify many of them depending on your needs. Unless you have a specific plan, if you want to scale any project up or down, make sure you change the ratios uniformly for any measured item. Double everything to make twice as much, for example.
- **Directions.** Instructions on how to make the project.

Safety

Many of the projects in this book require you to work with tools, heat, or other potentially dangerous things. Always wear the appropriate protective equipment and make sure other people and pets are at a safe distance.

The Three Rs

The three Rs stand for "reduce, reuse, and recycle." Together, they promote conservation and a sustainable living mindset. Apply the three Rs to things and energy.

- **Reduce.** Be mindful of what you currently use and find ways to cut down on your consumption. Use energy-saving appliances, for example.
- **Reuse.** Instead of throwing stuff out and/or buying new stuff, reuse it, either for its original purpose or another one.
- **Recycle.** Recycle whatever you can't reuse.

Going Off-Grid

Although it is not this book's primary intention, you can scale the projects listed in it to go completely off-grid. Here is some advice if you wish to do that:

- **Make a solid plan.** Do some further research and use what makes sense for your climate/country.

- **Plan backup systems.** When you are off-grid, you want at least two ways to do each thing. If you don't have a backup system you will be in trouble when something goes wrong. For example, if you rely 100% on a well for water and your pump breaks, then you'll have no water. But if you have a well and a rain catchment system, then you will not need to go without while your pump is getting repaired.
- **Take it step by step.** Going completely off-grid is a big project. Following your plan step by step will make it less overwhelming. It will also spread out your costs and ease you into an off-grid lifestyle.
- **Do it legally.** Make sure everything you do is legal and up to government code.
- **Get help if you need it.** Whether it is labor, technical knowledge, or a combination of both, use a reputable contractor and make sure he has a clear understanding of what you want and when you want it done. Get several fixed-cost quotes for the project, and pay the bulk of the fee when it is completed.
- **Buy your own materials.** This prevents a contractor from adding a profit margin.

FOOD

This section is the biggest in the book, and is split into six sections of its own.

Here, you'll learn a variety of ways to produce your own food and how to preserve it for storage or sale. There are also several alternative cooking methods you can try out.

VEGETABLE GARDENING

Growing your own vegetables is the most accessible way of producing your own food. It's also much healthier and cheaper than getting them from the supermarket or your local organic farmer. They'll be as fresh as they get and, if you do it right, chemical-free.

You can also grow medicinal plants.

Approximately the first half of these gardening chapters are general information, useful no matter what type of garden you choose to have. The second half focuses on specific, efficient gardening methods.

EFFICIENT GARDENING

An efficient garden is one that produces the highest amount of high-quality crops in the space you have available. The information in this chapter will help you achieve that.

Planning

The first step in getting the most out of your garden is proper planning. Read through all the information in this chapter and the rest of the gardening section so you can devise the best plan for your circumstances.

Companion Planting

Grow plants together that don't compete for resources. Wikipedia has an extensive list of companion plants:

https://en.wikipedia.org/wiki/List_of_companion_plants

Garden Type

The type of garden you choose will depend on your circumstances. Consider things like:

- How much space you have
- How much time you want to spend on it
- The climate you live in
- What you want to grow

Read through the different types of gardens in this section and choose which one(s) you want.

Harvesting

Harvest your plants at the peak of their ripeness to get the most nutrients. Pick them carefully so they don't get damaged. Throw any plants that are no longer producing in the compost and replace them. Use, store, or sell what you have picked as soon as possible.

How Much to Grow

Knowing how many of each plant you want to grow prevents over-planting. Calculate how much you consume, as well as how much you want to store and/or sell.

Intercropping

Intercropping is when you plant fast- and slow-growing vegetables together. This alone can triple your yield.

Maximizing Space

Traditional row planting and following what is printed on seed packets are an inefficient use of space. Zig-zag or block planting allows you to fit more in than you can with straight rows, and a crowded bed means there is less room for weeds.

Base your spacing on the final size of the mature plants and space them that way in all directions. You can also select plants that grow to smaller sizes. Here is a good resource for spacing plants:

https://gardeninminutes.com/plant-spacing-chart-raised-bed-gardening

Mulch

Mulching your garden reduces weeds and conserves water. It also provides food for the plants and worms.

Start with 1/2cm (1/4in) of grass clippings and add more as the plants grow. 2.5cm (1in) is a good amount for small- to-medium plants, and 10+cm (4+in) is a good amount for larger plants.

You can use other organic mulch, but avoid anything with seeds.

Plant Protection

The two biggest dangers to your plants are the weather and animals. Using either a single large greenhouses or several small individual ones will protect them from both. It will also extend the growing season in cold climates.

You can also introduce other plants and animals to combat the bad ones. The *Natural Insecticides and Herbicides* chapter has examples.

Rotate Crops

Different plants take and add different nutrients from and to the soil. Once one type of crop is done, plant a different type in its place to balance the soil naturally. Rotate them as follows:

Soil

Soil is one of the most important things for producing quality crops, and if you do it right, you'll never need to add any chemical balancers.

Use sandy loam mixed with compost in a 50/50 ratio as your go-to soil. Adding mulch on top will minimize weeds and release additional nutrients. You can also top it up with half an inch of compost every few months. Different gardening methods may require slight changes.

You can turn any soil into sandy loam by adding organic matter, though the process may take years depending on its current state. If you have poor-quality soil on your land, start adding organic matter to it now. While you're waiting, you can buy sandy loam to get your garden started. Within a season or two, you'll never have to buy soil or fertilizer again.

You can test if your soil is good by picking it up and forming a ball with it. Good soil forms a soft ball that will crumble if you press it with your finger. Soil that is too dry will break apart easily.

Another way to test your soil is by half filling a glass jar with it. Add water to the 3/4 mark, put a lid on it, and shake it up so there are no clumps. Leave it settle for one day. Once it has settled, it will be in three layers: sand, silt, and clay, in that order (sand on the bottom). Good soil has equal parts sand and silt and half as much clay (that is, 40% sand, 40% silt, and 20% clay).

Sunlight

Place your garden where it will get six to eight hours of sunlight a day. Orienting it from north to south will prevent shading.

Timing

Certain plants will thrive (or die) at different times of the year. Research what you plan to grow to see when the best time to plant it is.

You also want to have a constant crop so there is always something to harvest. This is called succession planting. To take advantage of succession planting, stagger the times you plant things and replant immediately once you harvest something. For example, start with cool-weather plants and once you get a crop, switch some of them out for warm weather plants.

In warm climates, you can grow all year round. To maximize growing times in cold climates, you need to consider the average frost dates. Every type of plant has different requirements, so check the seed packet or look it up.

Aim to have no free space throughout the season. Plant different plants in different places depending on the weather and their harvest times.

Tracking

Tracking your plants and what you do in a journal will allow you to refine your methods over time. After a few seasons, you will have a very efficient garden. Make sure to label or map your crops so you know exactly what is where and when you planted it.

What to Grow

Choose vegetables that grow easily in your climate and that you enjoy eating. If you plan to produce a lot, consider vegetables you can preserve sustainably and/or sell locally.

Weeding

Mulching will keep most weeds at bay, but some may still come through. Rip them out ASAP so they don't seed and spread. Weeding once a week will make it an easy job.

End of Season

At the end of each season, remove all the old plants and mulch everything with leaves. When the new season starts:

- Remove the leaves
- Mix in compost
- Rake the bed
- Plant the new batch

Related Chapters:

- Natural Insecticides & Herbicides
- Greenhouses

COMPOSTING

Composting is DIY fertilizer. It reuses all your organic scraps and provides a high-quality food for your plants. Unlike other fertilizers, compost will never burn your plants, so you don't have to worry about using too much. Composting is easy to do, but it takes time.

Location

You can make your compost heap in any dedicated area that's large enough. Many people like to put it in something such as a large bin or raised bed, but on straight on the earth is fine too. If you have minimal room (if you're in an apartment, for example), you can make a mini one out of a 20L (5gal) bucket with a lid and put it on your balcony.

When you're not using a container, 1m (3ft) cubed is a good size. You can make it wider or longer, but not any deeper. Any smaller and it will not heat up enough; any deeper and it will push the air out.

For convenience, put it somewhere within easy reach of a water source and your garden. Aesthetics are another thing to consider. You probably don't want a big pile of compost in the middle of your yard.

Wherever you choose to put it, make sure it's sheltered from the wind and sun. If you don't have natural protection (such as shade), you can cover it with straw or black plastic. When it's inside something, like a bucket, that doesn't matter.

Creation

Once you've decided where to put your compost heap, you can start to build it with the following alternating layers:

- 2 parts brown matter

- 1 part green matter
- 1 part kitchen matter

The ratio doesn't have to be perfect, but don't alter it too much. All the matter must be organic. Make sure there are no pesticides or other poisons in it. Big pieces of matter will work, but they will take longer to decompose.

The brown matter is carbon. Use brown grasses, hay, leaves, sawdust (from untreated wood), straw, shredded paper, chopped up cardboard, or almost any other dried organic matter. Spread it out evenly in your dedicated area and start to make it into a mound as it gets higher.

Green matter creates nitrogen. Green leaves, lawn clippings, green plant waste, and barnyard manure are all good.

Kitchen scraps include all your unused fruits and vegetables, as well as eggshells and coffee grounds. If you don't have enough kitchen scraps, use more green matter. Always cover food scraps with another layer of green matter to keep critters out.

Finally, top everything off with some ready-made compost, whether from a previous batch or store-bought. Alternatively, use any high-quality soil. It won't work as fast, but it will still work.

Soak the whole heap with water until it reaches the consistency of a wrung-out sponge, and then mix it all up. When you squeeze a handful of it, you should see one to five drops of water. Add more water if you need to. If it is too wet, add some more dry stuff and/or turn it daily to help it dry out faster.

Once it is all mixed, add some worms. This is optional, but recommended, especially if there is no way for them to get into the container naturally.

To speed the process up, cover the heap with black plastic, assuming it's not already in a covered container. This is more necessary in colder climates.

Some things that are bad for your compost heap include meat, domestic feces (dog, cat, or human), bones, fats, non-organic material (plastic, Styrofoam, etc.), poisonous or diseased matter, toxic materials, thorny branches, insecticides, and treated wood. Stay away from the leaves of black walnut, magnolia, hemlock, eucalyptus, juniper, pine, and oleander trees too.

Once a week, soak your compost until it reaches the consistency of a wrung-out sponge, and turn the pile every other day. Make sure you rotate the outside to inside. Mix in good soil occasionally just to give it an extra boost.

In one month, check to see if it's done. When it smells earthy and you can't recognize anything you put in it, it's ready.

Troubleshooting

- **No heat in the last 24 hours.** Add some water. In cold weather, the heap may just need a few more days.
- **Too wet.** Add dry stuff and turn it daily for a week or two.
- **Too dry.** Add water and cover it with black plastic or dry straw/hay.
- **Smells like ammonia.** Add more brown matter. If your ratio is good, then it's just not done yet.
- **Flies.** Turn it more often or cover it completely with straw and soil.

COMPOST TEA

Compost tea is full of nutrients and will protect your plants from diseases and insect infestations. Water your plants with it regularly except within the last four weeks before harvesting. You can keep some in the fridge for emergency spraying in case of disease.

What You Need

- 150L (40gal) plastic bucket with lid
- 4.5kg (10lbs) mature compost
- Water
- Long stick to mix

Directions

Place the bucket somewhere protected from extreme temperatures and close to a water source. Add the compost and fill it up with water. Stir it daily for a week. Strain the liquid and use it within 48 hours. It should not be overly smelly.

DRIP IRRIGATION

Drip irrigation is one of the most economical watering methods there are. It uses less than 10% of the water common sprinklers do and is healthier for your garden.

It works by dripping small amounts of water directly into the soil via piping.

What You Need

- Commercial-grade flexible polyethylene pipe
- Self-cleaning pressure compensating emitters
- Pipe fittings
- Rubber hose
- Water timer (optional)
- Pump (optional)

Quality tubing is important. Don't skimp on it. All hoses and piping must be safe to use for drinking water.

Directions

First, plan out where you will put the drip system. Measure it out and draw it up so you know exactly what you need to buy.

How you lay out the polyethylene pipe will depend on the layout of your garden, but do it so you use the least number of fittings as possible.

Once you know the layout of the pipe, you can figure out what fittings you need. Slip-ons are best. They go over the drip line, and friction keeps them in place. Ensure they are the correct size—they must fit exactly. Use t-fittings to direct water down a midline and a figure-8 end-stop to finish the line. The figure-8 holds the pipe in a folded position.

Emitter spacings ensure all the plants get the same amount of water. Self-cleaning ones prevent clogging. The number you need depends on the type of soil and the length of your pipe. For clay heavy soil, put one every 45cm (18in). In sandy soil, use one every 30cm (1ft).

The rubber hose connects the drip system (the polyethylene pipe) to your main water supply, so make sure you get enough of it to do that.

Attaching a water timer at the water supply point will save more water. It is especially useful when you're using an off-grid catchment system. Use the water in short bursts so your water catchment system has time to replenish. Water for 30 minutes, rest for 30, water for 30, rest, etc.

When you're using a sustainable water supply such as a well, you need to attach a pump. A 1/2 hp submersible pump is good for two acres. Make sure it has an automatic off switch to prevent it from burning out if the water supply runs dry.

HARVESTING AND STORING SEEDS

Saving seeds from plants so you can grow them in the future is the ultimate in sustainable gardening. Different seeds require different strategies, but there are some general guidelines you can follow, which are what you will learn here.

These apply when you're saving seeds from plants you have grown yourself. Don't try saving seeds from store bought fruits and vegetables—they probably won't grow.

Seed Collection

There are two types of seeds you can grow plants from: heirloom and hybrid.

Heirloom seeds produce plants with the same characteristics as the parent. Hybrids are more unpredictable, because they can take on characteristics from prior generations. It's best to save/use heirloom seeds, but hybrids still work. If using hybrids, prune out weak plants as you're growing them.

Regardless of whether they are heirloom or hybrid seeds, the best ones to save are those from the ripe fruits and vegetables of the plants that produce best. Let them mature on the plant before harvesting, which will be later than when you usually harvest.

Seed Preparation

Once you have your seeds, you need to clean and dry them. If you try to store them wet, they will spoil. Remove any pulp and spread them out on paper or a tight mesh screen. They need plenty of airflow. Leave them out for a few weeks.

Slippery seeds such as those from cucumbers, melons, and tomatoes must be fermented before you dry them. This is only if you plan to store them, not if you will grow them straightaway.

To ferment the seeds, put them in a container and cover them with water. Place the container somewhere warm, but not in direct sunlight. Stir them daily. They will be smelly. When the seeds drop to the bottom of the container, they are ready. Rinse them in water until they are clean and set them out to dry as normal.

Storage

Placing the seeds in an envelope is fine for short-term storage, up to 18 months. For long-term storage, put them in an airtight jar with a few desiccants.

Label the envelope or jar with the type of seed, the date of storage, and planting instructions. Place it in a cool, dark place.

Different seeds need different treatment. For example, some require stratification or soaking before planting. Research each type of seed you want to store and record the relevant instructions.

Although seeds can last in storage for many years, the success rate of germination drops over time.

Seeds from Herbs

To collect seeds from herbs, allow the herbs to flower. As the flowers die, cut the seed heads off. Put the seeds in a paper bag until they are dark brown/black. At this stage, they are dry and you can store them as normal.

Seeds from Peas and Beans

Allow pea and bean seeds to start drying on the plant. Pick them as they start to open and store them as normal.

REGROW VEGETABLES FROM SCRAPS

It is possible to regrow some vegetables, such as lettuce, herbs, and celery, from the scraps you don't use.

To do it, keep 5cm (2in) of the base intact. If you can, keep some of the roots too (in the case of spring onions, for example). Put this base in a glass with a few centimeters (1in) of water, and place the glass in the sunlight. Change the water every day. Once roots start to grow, transplant it to soil.

Onions and garlic can go straight into the soil. For garlic, the outer cloves work best. Plant them with the base facing down.

NATURAL INSECTICIDES AND HERBICIDES

Weeds, insects, and diseases can wreak havoc on your garden. Using high-quality soil, mulch, and compost tea will minimize these things, but sometimes they will still appear.

Here are some things you can do combat them without using chemicals.

Diseases

Compost tea is the best natural plant medicine there is. Use it regularly during watering for prevention, and spray it on the leaves if you see signs of disease.

Insecticides

An easy way to get rid of a bug you see on your plant is to blast it with a burst of water. Once it is gone, rub the plant stems with a cloth to destroy any eggs.

To attract and trap snails and slugs, leave out saucers of beer.

A simple homemade garlic spray is a good insecticide and fungicide. Crush several cloves of garlic and mix them with 1/2 cup of vegetable oil. Add that mixture to 3.5L (1gal) of water. Spray it on your plants, but don't use too much. To deal with larger pests, add chili to the mixture.

For a more permanent and preventative solution, use insect-repelling plants as a border garden and/or as companion plants. Many common herbs are perfect for this, and some of them will also repel annoying insects such as flies or mosquitoes.

These herbs include:

- Basil (flies and mosquitoes)

- Bay leaves (flies)
- Chives (carrot flies, Japanese beetle, aphids)
- Dill (aphids, squash bugs, spider mites, cabbage loopers, tomato hornworms)
- Fennel (aphids, slugs, snails)
- Lavender (moths, fleas, flies, mosquitoes)
- Lemon balm (mosquitoes)
- Lemongrass (mosquitoes)
- Lemon thyme (mosquitoes)
- Mint (mosquitoes)
- Oregano (multiple harmful insects)
- Parsley (asparagus beetles)
- Rosemary (multiple harmful insects)
- Thyme (whiteflies, cabbage loopers, cabbage maggots, corn-ear worms, whiteflies, tomato hornworms, small whites)

There are also several ornamental plants that have similar properties, such as chrysanthemums.

Animals are also good for preventing bugs. Create an environment that attract animals that will eat them. Birds eat numerous types of insects, for example, while ladybugs eat aphids.

Herbicides

This natural herbicide will kill weeds, but also affect other plants, so be careful of what you spray it on. To make it, you'll need:

- 3.5L (1gal) of white vinegar with a maximum of 5% acidity
- 1 cup of salt
- 1 tablespoon of liquid dishwashing soap (optional)

Mix the salt and vinegar together and stir it until the salt dissolves. Add the dishwashing liquid if you want. It will help the solution to stick to weeds that have a protective coating, but is not organic.

Put the mixture in a spray bottle (or make it directly in the spray bottle) and spray it on the weeds.

Related Chapters:

- Compost Tea

GREENHOUSES

A greenhouse is a transparent house for your plant(s). Its main purpose is climate control (which can extend your growing season), but depending on how you set it up, it can also keep your plant(s) protected from insects and critters.

There are many ways to build a greenhouse. The main things to consider are its size, permanency, and aesthetics.

A miniature greenhouse is a cold frame. You can make one out of old windows and scrap wood.

A hoop house is a good option for a medium-sized greenhouse. It is a series of large hoops (made from PVC piping, for example) covered with greenhouse plastic. You can also make a smaller version, which can be easier to construct than a cold frame.

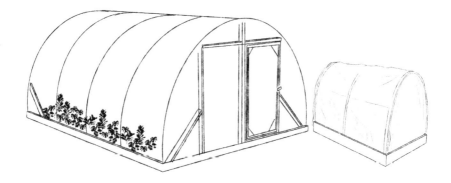

Traditional greenhouses are larger, permanent structures made from wood and glass.

In larger greenhouses, you need to pay extra attention to climate control. When one is over-heating, open the door. If it gets too cold, consider installing heat lamps and/or an electric heater.

There are many free designs on the internet for all the above types of greenhouses. For this project you will make a simple individual bottle greenhouse.

Get a large, clear plastic bottle. Cut the bottom off, and remove the lid. Put it over the plant.

GARDEN BEDS

You can turn any patch of dirt into a garden bed. Be sure to use the tips in the efficient gardening chapter to get the most out of it.

What You Need

- Space to put it in
- A shovel
- Various types of organic matter
- Seeds or plants to plant

Directions

First, plot where you want to put your garden bed(s). The size of an average door is a good size for each bed. If you plan to have multiple beds, make sure you leave some space to walk between them.

Take off the topsoil and put it aside. It is roughly the top 20cm (8in). With the topsoil saved, dig the bed out an additional 30cm (1ft), so the hole is 50cm (20in) deep in total. Remove all the stones, weeds, twigs, and other non-dirt matter from it.

Fill the hole with the following, in order from bottom to top:

- 5cm (2in) solid organics (eggshells, bones, wood, paper, rocks, etc.). This layer helps with drainage.
- 20cm (8in) of mulch (grass, seedless hay, etc.)
- The soil you dug out
- The topsoil, mixed with compost.

It will be about 15cm (6in) above ground level.

Fence it off to protect is from any animals (or children). For example, make a border out of stones, or build a fence and grow a creeper on it.

Leave the bed for one week before planting. When it is time to plant, water the bed well and allow it to drain so it's not muddy. Plant your seeds (or plants) and press the soil down firmly on them. Sprinkle them lightly with water. A plastic bottle or tin with small holes punched in it makes a good improvised watering can.

Water as needed depending on what you are growing, but preferably in the morning, especially in hot areas.

Fruit Trees

To plant a fruit tree, dig a hole $1m^3$ and fill it with organic matter as before, but leave room to plant the tree. Soak the hole with water and let it drain. Plant the tree so the soil around it is level with ground. Stomp around it to make it firm, then water it. Splint the tree to a strong stick to protect it from strong winds or other things if needed.

Related Chapters:

- Efficient Gardening

RAISED-BED GARDEN

A raised-bed garden is like a traditional garden bed, but it is elevated. It has several advantages over a tradition garden bed:

- Better drainage
- Deeper rooting
- Improved soil
- Increases the growing season
- Requires less bending and kneeling
- Reduces diseases
- Allows for more growth in small spaces
- Reduced weeding

The downside is that the soil will dry faster, so you need to water it daily or install a drip irrigation system. The latter is recommended.

Combining raised beds with vertical gardening can give you a large yield in a relatively small area.

What You Need

- Space to put the bed in
- A shovel
- Scrap materials to make a wall around the bed
- Soil
- Compost
- Seeds or plants to plant
- Weed fabric (optional)

Directions

Placement is your first consideration. The bed needs to be about 1m (3ft) wide and however long you want it. You also need to make sure it will get six to eight hours of sunlight every day.

To prepare the site, loosen up the first 1m (3ft) of depth in your plot. To make grass easier to remove, kill it off by covering it with cardboard or plastic for six weeks.

Next, build a wall around the bed area you have marked out. You can use planks of wood, bricks, cinderblocks, etc. Make it at least 1/2m (1.5ft) high. For improved accessibility, make it waist high.

If you want, put weed fabric down. This will block weeds while still allowing for drainage.

Fill the bed with sandy loam mixed with compost in a 50/50 ratio. To ensure the soil has no seeds, buy it new. Mounding the soil will increase growing space.

Plant your seeds/plants as normal.

In hot climates, adding mulch on top will help to retain moisture.

Related Chapters:

- Drip Irrigation
- Vertical Garden

VERTICAL GARDEN

A vertical garden is one in which plants grow up and off the ground. It maximizes space, and is more pest- and disease-resistant than most other types of gardens. It also requires minimal maintenance, through you need to check it for weeds often, as there is no room for competition.

Combining vertical gardening with a raised-bed garden and a drip irrigation system is an ideal example of high-efficiency gardening.

You can grow almost anything vertically if you really want to, but different plants have different requirements. Some of the best edible plants to grow in a vertical garden include:

- Broccoli
- Carrots
- Cauliflower
- Cucumbers
- Garlic
- Herbs (basil, chives, cilantro, mint, oregano, parsley, rosemary)
- Leafy greens (kale, lettuce, spinach)
- Peppers
- Strawberries
- Tomatoes

There are many things you can use to create your vegetable garden. Some examples include:

- Arches
- Baskets
- Chicken wire
- Cinderblocks
- Containers
- Fences

- Gutters
- Long poles
- Pots
- Trellises

Search "vertical garden images" on the internet for inspiration.

When choosing the design and what to grow, consider drainage and the weight of the plants and wet soil. Take care not to block sunshine to any crops growing underneath the garden.

Here is an easy and cheap way to make a vertical garden. It uses empty plastic bottles, but you can adapt it to almost any container you have lying around.

What You Need

- Empty plastic bottles
- A sharp knife and/or scissors
- Twine or wire

Directions

Place the bottle on its side and cut a rectangular hole in it. This side is the top.

Create a small hole in the center of the short side of the rectangular hole. Do the same thing on the other side, so the two holes line up horizontally.

Make two more small holes on the bottom of the bottle, so they line up with the small holes on the top. Put one more small hole on the bottom in the middle of the rectangle. This is for drainage.

Thread the twine through the small holes on either side of the rectangle. Tie a knot in the twine underneath the bottom holes where you want the bottle to sit. Alternatively, tie it around a washer or other small object. This stops the bottle from slipping.

Repeat this for as many bottles as you want.

Tie the tops of the twine wherever you want to put your hanging garden. Make sure it will get six to eight hours of sunlight.

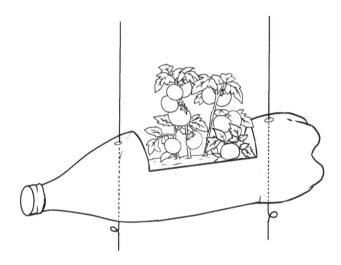

Related Chapters:

- Drip Irrigation
- Raised-Bed Garden

THREE SISTERS GARDEN PLAN

The "three sisters" are corn, beans, and squash. Growing these together provides your body with carbohydrates (corn), protein (beans), vitamins and seed oil (squash), and maximizes space while doing so.

The corn grows in the middle of the plot. The beans climb up the corn stalks and the squash surrounds both of them on the ground below.

This method of growing provides a good yield and is low maintenance, since the three plants support each other. The beans fertilize the ground, the squash leaves and other matter act as a self-composter, and the squash spines keep pests away.

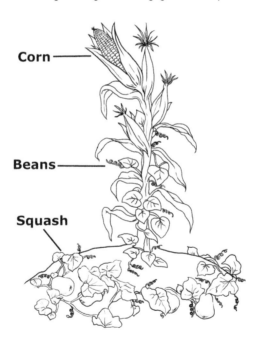

What You Need

- A mound of soil 30cm (1ft) high and 1.2m (4ft) wide/long.

- Corn seeds.
- Runner or pole bean seeds.
- Squash/pumpkin seeds. Use a variety with cascading vines, as opposed to a compact bush.

Directions

One day before you plan to plant (depending on the average frost times), soak the mound in water. Twenty-four hours later, plant six corn kernels 2.5cm (1in) deep and 25cm (10in) apart in a circle 1/2m (2ft) in diameter. Add water to the top of the mound.

Once the corn is 10cm (4in) tall, plant six bean seeds spaced evenly around each stalk. Add water to the top of the mound. One week later, plant three squash seeds evenly spaced around perimeter of mound. Add water to the top of the mound.

Once all the plants are sprouting, weed out any weak ones. If the beans aren't naturally climbing the corn stalks, wind them around the stalks to train them.

Always water the crop from the top of the mound, as opposed to directly on the leaves.

PERMANENT MULCH GARDENING

A permanent mulch garden (also called the Ruth Stout or "no-work garden") is one of the lowest-maintenance gardens there is. All you have to do is cover your land in mulch once a year, plant the seeds, and water it.

This style of gardening produces rich soil, increases moisture retention, and eliminates weeds. It also protects against erosion, so you can do it on sloped surfaces.

With a permanent mulch garden, there is no need for a separate compost pile. Instead, you can just place your food scraps underneath the mulch.

What You Need

- A plot of land
- Mulch

Your mulch can be almost any non-toxic organic matter that rots, such as straw, hay, leaves, or wood chips. Use what it easy and cheap (or free) to get in your area.

Planting mulch-creating trees (hornbeam, beech, oak, and others) around your property will also provide you with a windbreak and natural security.

Directions

Start covering your plot of land with mulch as soon as possible. The longer it has to break down before planting, the better. If you cover it in autumn, you can seed in spring. Cover it with mulch at least 20cm (8in) thick. If you use lighter mulch, such as hay, lay it on thicker, since it will settle. Do not mix the mulch into the soil.

When it is time to plant, pull the mulch away from where you want to plant the seed. Plant the seed and cover it with dirt, but not mulch. As it grows, pull the mulch back a little more.

Each season, choose a different place to put your plot, and leave old areas covered in mulch for a couple of years. If you add new mulch to each plot every year, all your soil will be excellent in a few years.

To convert a traditional garden into a permanent mulch garden, cover the earth around your plants in mulch.

Troubleshooting

- **Weeds**. Pull them out and add more mulch.
- **Snails and slugs**. Pull the mulch back from the plants a bit more.
- **Pale leaves**. Add grass clippings and/or coffee grounds to increase the nitrogen.

KEYHOLE GARDEN

Keyhole gardens were first developed in Africa to maximize water efficiency when growing vegetables. They are good to use in dry areas.

The structure of this kind of garden resembles a keyhole, hence the name. It is a circular raised bed garden with a compost pile in the middle. The "keyhole" gives you access to the compost pile.

What You Need

- A plot of circular, flat land 2m (6.5ft) in diameter
- Chicken wire
- Stakes
- Rocks
- Composting materials
- Longer poles (optional)
- Shade cloth (optional)
- Scrap materials to make a retaining wall
- Soil

Directions

Clear your chosen plot of land, then build a compost basket in the middle of it using the chicken wire and stakes. Put the rocks at the bottom of your compost basket for drainage, followed by a layer of topsoil. Fill the rest of the basket with composting materials.

If you want, use the longer poles and shade cloth to construct a roof. This will protect the compost pile from drying out or over-soaking during rain.

Build a retaining wall at the edge of the plot of land 1m (3ft) high. Use whatever you have (bricks, rocks, wood, etc.). Leave a path so you can access the compost pile.

Layer soil between the wall and the compost as a garden bed, as described in the Garden Beds chapter. Make it slope away from the center. Let it sit for one week before planting.

When your first plant, you will need to water the soil directly. Once the roots are established, the water from the compost will sustain the garden.

Continue to top up your compost heap and maintain the garden bed as normal (weeding, bug control, etc.)

Check out designs on the internet for inspiration.

Related Chapters:

- Composting
- Garden Beds

HYDROPONICS

Hydroponics is the practice of growing plants in mineral nutrient water. It is a little more complicated to set up and maintain than soil gardening, but it has many advantages. For example:

- You can do it indoors with lights (though it is better outside with sunlight)
- It's efficient (faster and bigger yields)
- There are fewer diseases and pests
- It's low- maintenance
- There are no weeds
- It saves water

The biggest disadvantage of hydroponics (arguably) is its reliance on electricity and specialty equipment such as a pH tester and a pump. You can make it sustainable by running the pump off solar power.

There are several ways to go about creating a hydroponic garden. The main difference between them is how they deliver the water, oxygen, and nutrients. The easiest structure to construct and maintain for most people is deep water culture (DWC), so this is the method these instructions describe. If you want to learn about the other methods, research the following:

- Ebb and flow (or flood and drain)
- Wicking
- Drip
- Aeroponic
- Nutrient film technique (NFT)

With a deep water culture hydroponic setup, the plant roots are suspended in nutrient-rich water oxygenated by a pump. The following DWC setup is good for growing small to medium-sized plants, but you can scale it up if you want.

What You Need

- 20L (5gal) bucket with lid.
- Black and glossy white paint. Must be good for plastic.
- Plant basket.
- Air stone. Buy a weighted one so it doesn't float.
- 1/4in airline.
- Air pump.
- Growing medium. Expandable clay pellets or diatomite are best.
- Nutrient mixture.
- Plants.
- pH tester (either strips, liquid, or digital).
- pH up and pH down solutions.

You can purchase all of the above from any decent hydroponics store or online.

Directions

Cut a notch in the rim of the bucket about 5cm (2in) down and a couple of cm wide (1in). This is for the airline.

Paint the outside of your bucket black to make it light-proof. Once it's dry, give it a coat of glossy white so it reflects heat.

Cut a hole in the bucket lid so your basket can sit in it. Do it in a piece of cardboard first and use that as a template. Using a rotary tool, if you have one, will make this easier.

Fix the air stone to the bottom of the bucket, then run the air line from the air stone to the air pump outside the bucket.

Place your bucket where it will get a minimum of six hours sunlight each day. Alternatively, you can set it up indoors with artificial lights. If this is your plan, you'll need to do more research depending on what you want to grow.

Once your bucket is in place, fill it up with at least drinking-grade water. The higher quality the water, the better. Reverse osmosis is good. Pour the water until the bottom of your basket is 2cm (3/4in) underwater. Mark this water line with a permanent marker all the way around. This will give you a good reminder of how high to keep the water.

Mix the nutrient solution in the water according to the manufacturer's directions. Stir it and wait one hour, then check and adjust the pH as per the instructions at the end of this chapter.

Fill the basket with growing media and your seedling or plant.

Using an existing plant is easiest. Submerge the roots in fresh water and gently wash them with your hand. Make sure they're completely clean, so you don't contaminate the water.

Alternatively, start seedlings in starter cubes, then transplant them into the growing media. Make sure your water level is high enough that the root balls get plenty of moisture.

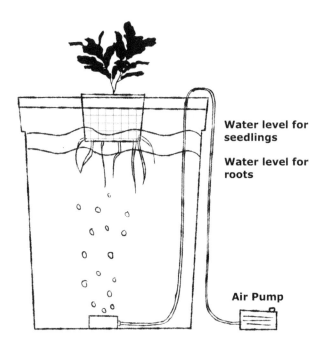

Maintaining pH Levels

Once you are set up, you need to check and maintain the water's pH level daily.

To test the water, follow the instructions for the type of pH tester you have (strips, liquid, or digital).

Most plants prefer a pH level between 5.0 and 6.0. Research the optimal level for the plants you are growing. Adjust the water's pH with the pH up and pH down solutions. Add the solutions in small doses to avoid overcompensation. Stir them in, then leave the mixture for one hour. Retest and readjust it as needed.

AQUAPONICS

Aquaponics is aquaculture and hydroponics combined. With it, you grow fish and plants at the same time, and they complement each other's growth in a circular system. The fish poop feeds the plants, and the plants clean the water for the fish. All you have to do is feed the fish.

Following these instructions, you will create a small, stacked", media-based aquaponics system. It is ideal for the average home.

The plants are grown in a grow medium inside a grow bed, and the grow bed sits on top of the fish tank. The water is pumped from the fish tank to the grow bed. It passes through the grow medium and nourishes the plants. The plants clean the water and it return to the fish tank by gravity.

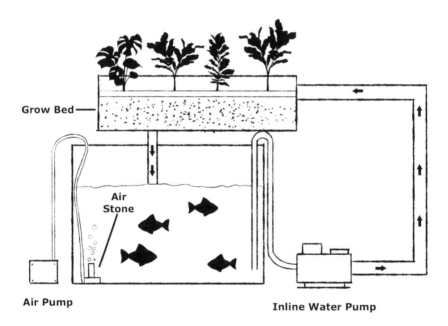

What You Need

- 2 food-grade plastic containers. One is the fish tank and one is the grow bed. They can be the same size, but if one is bigger, make that the fish tank.
- Gravel for the fish tank. Use a natural color.
- Air pump.
- Air stone.
- 1/4in air tube, long enough to connect the air pump to the air stone.
- Mesh.
- Inline water pump. You could use a submersible pump instead, but would need to set it up a little differently than in this design. An inline pump system allows for greater expansion later, as it is more powerful. The size of the pump depends on the amount of water it needs to circulate. The pump salesperson can help you choose the correct size for your system.
- Enough water tubing to feed water into the grow bed.
- Grow medium. Choose 1/2in or 8mm to 12 mm expandable clay pellets.
- Aquarium water test kit(s) and balancing solutions.
- Fish. You won't need the fish straightaway.
- Plants. You won't need them straight away.
- Timer (optional).

Some choices for fish:

- **Goldfish.** Cheap to buy and easy to maintain, but not good to eat.
- **Tilapia**. Good for warmer climates. Easy to grow and good to eat.
- **Trout**. Good for colder climates, around 20C (68F).

Some choices for plants are:

- Chilies
- Cucumber
- Eggplants
- Leafy greens
- Melons
- Tomatoes

Many other plants and fish will also work. Research what you like to eat.

Directions

Review the instructions for the hydroponic setup for a better understanding.

Work out where you want to place your system, because it will be hard to move once it's set up.

Fix the air stone to the bottom of the fish tank and attach it to the air pump with the 1/4in air tube. Wash the gravel and spread it evenly on the bottom of fish tank. You only need a thin layer. Fill the fish tank up with water and turn on the air pump to start oxygenating the water.

Test and balance the pH, ammonia, nitrites, and nitrate levels. Ammonia and nitrite levels should always be non-detectable. Nitrites should be low, and the pH should be neutral (6.8 to 7.3).

Once the water has been treated, leave the tank with the pump running for at least one week before adding the fish.

Punch several small holes in the bottom corner of your grow bed so water can drip back into the fish tank. Cover the holes with several layers of mesh so the growing medium won't block them.

Place the grow bed on top of the fish tank. Make sure you leave a gap so you can access the fish tank while staying balanced. Clean the growing medium and fill up the grow bed with it, leaving 10cm (4in) of space from the top. Make sure it's spread evenly.

Connect tubing from the bottom of the fish tank to the pump, and then into the side of the grow bed. Turn on the pump to test the system. You can also connect the pump to a timer so it will automatically switch on and off when you want it to (30 minutes on and 30 minutes off, for example), but running it non-stop is also okay.

When the system is ready (at least one week after the fish tank setup), add fish to the fish tank. Only add one or two fish at a time, and wait a week before adding more.

Add the plants to the grow bed by placing the roots gently in the grow medium. Ensure the roots are deep enough to draw nutrients from the water. If you want to grow seeds, make sure you have at least a couple of plants with good roots too, to help with water filtration.

Maintenance

Feed your fish a quality diet using flakes or pellets. Do not give them anything live, as that may introduce disease. Feed them two times a day, preferably at the same time every day. Give them only what they can eat in five minutes.

Test and balance the fish water weekly. If the system is working well, it shouldn't need much (if any) adjustment.

Tend to the plants as normal.

Do further research about caring for your fish depending on the type of fish you have.

Related Chapters:

- Hydroponics

SOY

Soybeans are versatile, sustainable, and very nutritious. Many vegetarians eat soy in various forms as their main meat substitute, but even if you're not vegetarian, the beans are great to add to your diet.

There are three main types of soybeans so make sure you choose the right one for what you want. The black and green ones, like Japanese edamame are what you can cook and eat. The yellow ones are good for most of the projects in this section (making flour, milk, oil, etc.).

Soybeans grow like any other plant. When the pods are green and 5cm (2in) to 8cm(3in) long, they are ready to harvest. Pick the entire pods, then blanch and shock them.

To do so, prepare two pots of water. One is for boiling water and one is for iced water. Put the pods in the boiling water for five minutes. Remove and strain the beans, then put them straight into the iced water. After they have sat in iced water for five minutes, remove the beans from their pods. You can compost the hulls or feed them to your chickens.

If you don't want to grow your own soybeans, you can buy them. For the best results, choose high-quality, non-GMO, organic beans. Pick out the bad ones and wash them immediately before using them.

Quality soybeans will have a clear hilum (the little line on edge) and will be uniform in size.

When buying dried beans, you may need to soak them, depending on what you want to do with them. Soak them overnight (preferably for 24 hours) in three times as much water as there are beans. Changing the water every eight hours or so will prevent bacteria from growing.

If you have fresh soybeans and you want to dry them (to make soy flour, for example) boil them for 25 minutes, then drain and rinse them. Spread them in single layer on a baking tray and roast them in the oven at 110C (230F) until they're completely dry. It will take about 90 minutes.

Other options for drying boiled soybeans include dry-roasting them on a medium-to-low flame for about five minutes or sun-drying them for at least two days.

OIL

You can do a lot of things with oil and making it yourself is cheaper, healthier, and tastes better than buying it. Some of the things you can use your homemade oil for are:

- Biofuel
- Candles
- Carrier oils for essential oils
- Cooking
- General lubricant
- Oil lamps
- Soap

Although it is in the soybean section, this chapter will teach you how to make oils in general, including those from soybeans.

Oil Types

There are many types of oils you can make depending on what you want to grow (or buy).

Sunflowers are easy to grow, and their seeds make a good all-purpose oil. Use the black seeds to make the oil. The ones prepared for humans to eat are not good for making oil, and you cannot grow them.

The seeds you want to make the oils from need to be "just right" before they can be picked. If they're too moist, they won't produce enough oil. Too dry, and they'll clog the oil press.

You will also need to hull or husk them before pressing. You can do it by hand or use a seed mill.

Things like almond, pumpkin, hazelnut, and coconut make fragrant and flavorsome oils. They are good for cooking (depending on what you are cooking) and for making soap, candles, and other things.

Different oils have different properties, which means they have different yields, shelf lives, and production processes. Make sure you research the oils you want to make to find out the specifics.

Cold-Pressing

There are two ways to extract oil: hot-pressing and cold-pressing.

Cold-pressing is easier and produces better-tasting oils, but it won't work for everything. Some oils that are good for cold-pressing are canola, olive, sesame, soybean, and sunflower.

You need an oil press to do cold-pressing. When buying one, you can choose between manual or electric, and whether or not you want it to work as a hot press too. Whatever features you choose, make sure you get a reputable brand, such as Piteba or Forkwin. Clean your press after each use.

The basic method for cold-pressing is to hull the seeds/nuts (hulling is optional for sunflower seeds) and put them through the press. Do multiple pressings to get more oil. "Virgin" oils are from the first press and are the most flavorsome. Throw the seed pulp in the compost or feed it to your chickens.

To cold-press olive oil, get some fresh black olives. You can use green olives too, but black ones will give you a higher yield. Wash and pit them, then mash or blend them into a paste. Stir the paste slowly for 45 minutes to release the oil, and put it through the press.

You can use an orange press to cold-press avocado oil. That way, you still get to eat the avocado. Halve the avocado and remove the seed and flesh. Put the skin through the orange press to extract the oil.

Hot-Pressing

Hot-pressing works with more varieties of oil, and you get a higher yield. You can also do it without an oil press.

To hot-press without using a press, first hull the seeds/nuts and then dry them, either in the sun or the oven. Be careful not to over-dry them. Research to determine specific drying times for the seed or nut you want to dry.

Once they're dried, mash them into a pulp and boil the pulp in a pot. Do not let it smoke. Collect the oil off the top or let it cool down and it will separate. Strain and store it.

If your press can hot-press, follow the manufacturer's instructions.

To hot-press coconut oil, husk the coconut and remove the flesh. Grate and dry the meat, then press it as previously described.

Storage

Put your oil in a dark glass container to store it. Place the container in a cool, dark, and dry area. Most oils keep for at least one year, but shelf lives vary.

Filtering the sediment out of your oil before you store it is optional. There are two ways to do so. The first is to let it sit for a few days, and syphon off the oil once the sediment has settled on the bottom. The other method is to strain the oil through a coffee filter or cheesecloth.

SOY FLOUR

Soy flour is a high-protein alternative to wheat flour. It is low-carb, wheat-free, and gluten-free.

Some recommended uses for soy flour are:

- Substituting up to 30% of other flours in any recipe.
- Thickening sauces and soups.
- Making flatbreads, such as roti, from 100% soy flour.

What You Need

- Dried yellow soy beans
- A grain mill or a good blender
- A sieve
- An airtight container

Directions

Grind or blend the soybeans until they are powder, then sift the powder into your airtight container. You can store soy flour for up to 12 months.

SOY MILK

Soy milk is easy to make and is healthy. You can drink it as you would any other milk, or turn it into soy yoghurt or tofu. There are two ways to make soy milk: from fresh soy beans or from soy flour.

What You Need (Fresh Method)

- Fresh yellow soy beans
- Drinking water
- A blender
- A nut-milk bag
- A pot
- A wooden spoon
- A sealed container
- Flavoring, such as vanilla beans, cocoa, or dates (optional)

Directions (Fresh Method)

Soak the fresh soybeans in water for 12+ hours. Strain, rinse, and remove the skins.

Blend them with water until the mixture is smooth. Use one part soybeans for every eight parts water. For example, use half a cup of soy beans and four cups of water.

Once the mixture is made, strain it through the nut-milk bag and into the pot. Squeeze out as much as you can. Bring the liquid to a gentle simmer, and keep it there for 20 minutes. Stir it often and remove any skin that forms on the top.

After 20 minutes, allow it to cool. Blend in any flavoring and store the milk in a sealed container. It will last up to four days in the fridge.

What You Need (Flour Method)

- Soy flour
- Drinking water
- A pot
- A stove
- A whisk
- A wooden spoon
- Cheesecloth
- A colander
- A bowl
- A sealed container
- Flavoring, such as vanilla beans, cocoa, or dates (optional)

Directions (Flour Method)

To make soy milk using the flour method, you need one part soy flour to three parts water (one cup of soy flour to three cups of water, for example).

Bring the water to a low boil, then slowly pour in the soy flour. Whisk it as you pour so it is completely mixed. Reduce the heat to a simmer and stir it constantly until you get the right consistency. This will take about 20 minutes. Once it's done, allow it to cool.

Put the cheesecloth in the colander and place the colander over the bowl. Pour the cool liquid over the cheesecloth. Squeeze out as much liquid as you can.

Blend in any flavoring and store the milk in a sealed container. It will last up to four days in the fridge.

Related Chapters:

- Soy Flour
- Soy Yoghurt
- Tofu

SOY YOGHURT

Homemade yoghurt is healthier than store-bought. There are no added preservatives or sugars, and it is not pasteurized, which means there are more healthy bacteria.

The process for making soy yoghurt is the same as the one for making normal yoghurt, and you can use the product in the same way too, whether in smoothies, curries, or plain with your favorite toppings.

What You Need

- Plain soy milk (soybeans and water)
- Starter culture (either yoghurt starter or some homemade soy yogurt from a previous batch)
- An oven
- A pot
- A wooden spoon
- Glass containers

Directions

Preheat your oven to 50C (120F). Heat the soy milk in a pot to about 90C (194F). Do not boil it. Pour it into the bowl and let it cool to 50C (120F). At 50C, it will be hot to the touch, but won't burn you.

Stir the starter culture in, then put the mixture into the glass container. Turn the oven off and put the jars inside. Leave them in there for 12 to 15 hours. Do not open the oven during this time. When the yoghurt is ready, cover the jars and refrigerate them.

If you don't have an oven, you can use a yoghurt maker. The main thing is to keep it at a consistent temperature of about 50C (120F) for 12 to 15 hours.

TOFU

Tofu is coagulated soy milk. The process to make it is like the one for making cheese, and the end product provides your body with protein, fats, carbs, essential amino acids, and a variety of vitamins and minerals.

Here are two ways to make tofu. One is using soy milk and the other with soy flour.

What You Need (Flour Method)

- 6 cups of water
- 2 cups of soy flour
- 6 teaspoons of vinegar or lemon juice
- A pot
- A sifter
- A whisk
- A wooden spoon
- A colander
- Cheesecloth
- A bowl
- A sealed container

Directions (Flour Method)

Bring the water to a boil then slowly sift in the flour, whisking continuously as you do so.

Continue to stir the mixture until it starts to boil again. Do not let it boil over. As soon as it boils, add the six teaspoons of vinegar or lemon juice. Add one teaspoon at a time, and stir slowly as you do so.

Once the mixture turns to curds, remove it from the heat.

Line the colander with the cheesecloth and place it over a bowl. Pour the curds into the cheesecloth and, if you like, run it under cold water to cool it.

Squeeze the moisture out of cheesecloth. This is also optional, but the less moisture it has, the firmer the tofu will be.

Shape the tofu however you want, or leave it "scrambled," and store it in a sealed container in the fridge.

What You Need (Milk Method)

- 2L soy milk (room temperature)
- A pot with a fitted lid
- A stove
- A wooden spoon
- Coagulant
- 1/2 cup water
- Cheesecloth
- A colander

Choices for coagulant:

- 1.5 teaspoons of nigari crystals
- 1.5 teaspoons of gypsum
- 1.5 teaspoons of Epsom salts
- 2 teaspoons of liquid nigari
- 75ml of lemon juice

Directions (Milk Method)

Pour the soy milk in the pot. Bring it to a boil with the pot lid on, then lower it to a simmer with the lid off. Simmer the milk for five minutes while stirring constantly. Remove any solids that form on the surface.

Once the five minutes is up, take the milk off the heat and let it sit for a few minutes. While waiting, mix the coagulant with half a cup of water. Stir it in so it dissolves.

You need to add the coagulant to the milk in three stages. As you stir the milk, add one third of the coagulant and then let the liquid settle. Once it has settled, sprinkle another one third of the coagulant in. Cover the pot, wait three minutes, then sprinkle the last one third of the coagulant in.

Gently move your wooden spatula on the top few cm (1in) of the mixture for 20 seconds. Give extra attention to any milky liquid on edges. The curds will begin to coagulate. At this stage, cover it for three minutes (six minutes if using gypsum or Epsom salts).

If it is still milky after the three (or six) minutes, stir the surface for 20 seconds. You want white chunks (curds) in pale yellow liquid (whey) with no milkiness, like cottage cheese.

If it still has liquid after the 20-second stirring, cover it for one minute, then gently stir the surface again. Put it over very low heat for a few minutes, gently stirring the surface. Turn off the heat, cover it, and leave it for two minutes.

If it still has liquid in it, add one quarter of a teaspoon of coagulant (or half a teaspoon of liquid nigari or 20ml of lemon juice) to one third of a cup of water and sprinkle it into the milky areas while gently stirring the surface.

Line the colander with the cheesecloth and put it in the sink. Put some whey (liquid) on the cheesecloth to moisten it. Remove as much of the rest of the whey as possible from the curds.

Gently put the curds into the cheesecloth, then fold the cloth over them. Weigh down the tofu with something that will weight it down evenly. The heavier the weight and the longer it's left on, the firmer the tofu will become. Using a one-kilo weight for 20 minutes will produce tofu of medium firmness.

The texture will become firmer as it cools. With experimentation, you can get the consistency you want.

When you are happy with it, submerge the colander in water, then take out the cloth with the tofu in it. Remove the cloth, cut the tofu, and submerge it in water for five minutes.

Put it in containers, cover it with water, seal it, and put it in the fridge.

Related Chapters:

- Soy Flour
- Soy Milk

MISO

Miso is a Japanese seasoning made by fermenting soybeans with various ingredients. It is hard to mess up, but it takes months to ferment. However, it is worth the wait, as it is very versatile. You can use miso:

- As a dipping sauce (mix it with mayonnaise)
- In salad dressings (search the internet for "miso salad dressings")
- As a marinade
- As a gravy
- In soup
- To replace salt
- To replace soy sauce (mix the miso with a little water)
- To thicken sauces, curries, or stew
- In vegetable stock

For best results, ferment the miso over a cold winter, a cool spring, and a hot summer. Start a couple of months before the weather turns warm, when it's about 10C (50F) outside. You also need to ferment it at about 27C (80F) for a few months.

If you do not live somewhere with seasons like this, you can simulate them. For example, in the tropics, you will put it in the freezer, then the fridge or a cool room, then room temperature.

The instructions below are for fermenting it through cold weather first, which will give the miso a deeper flavor, but you only need hot weather to ferment it.

If you make it in the summer, it may not get the deep flavor, but will be ready in four months (as opposed to six months). If the temperature is warm enough, you can leave it inside at room temperature.

This recipe makes 3.5kg (7.5lbs) of miso.

What You Need

- 1kg (2lbs) of dried yellow soybeans
- 2 teaspoons of miso paste
- 1kg (2lbs) of brown- or white-rice koji
- 400g (14oz) of fine, white sea salt
- Muslin cloths
- A pot
- A food-grade container
- A plank of wood that fits inside the container
- A weight to put on top of the wood

Directions

Soak the soybeans for at least 12 hours, then drain them and bring them to a boil in water that is about 10cm (4in) above the beans. Once they're at a boil, lower them to a simmer and leave them to simmer for about two hours. They are ready when you can easily crush them with your fingers. You want most of the liquid to be boiled off by the time the beans are cooked.

While waiting for the beans to cook, whisk half a cup of hot water with the two teaspoons of miso paste to make a very thin miso soup. You will use this like a starter culture for yoghurt. Allow it to cool to room temperature.

Mash the cooked beans to a consistency you like and let them cool to room temperature.

Add the rice koji, three quarters of the salt, and the thin miso soup. Mix it all together so it is evenly distributed.

Put the mixture into a large food-grade container. It should fill no more than half the container. You need to get out all the air from the mixture. Do this by making palm-sized balls and throwing them into the container. Once all the balls are in the container, pat the mixture down firmly, so that it's flat, and sprinkle the remainder of the salt on top.

Wipe the inside of the container above the mixture with vodka to prevent mold from forming, then place the clean muslin cloth over the mixture to protect it. On top of that, place the piece of wood and weight it evenly with an amount of weight that is at least equal to what the mixture weighs. Cover it with another muslin cloth and tie the cloth in place. It must be cloth and it will collect mold.

Assuming it is about 10C (50F) outside, put the container outside in an undisturbed shaded area. If not, put it in the fridge until the weather cools.

Leave it throughout winter. Ensure it is never in direct sunlight.

In the spring, start stirring it once a month. Carefully remove the cloth and wash it. Scrape off any mold and stir the mixture. Pat it down, then replace the cloth and weights.

In summer, stir it every two weeks. After summer, it will done, but you can leave it a little longer if you want more depth of flavor. Store it in sealed containers in the fridge. It will last at least one year.

Related Chapters:

- Soy Sauce
- Fermenting

SOY SAUCE

Soy sauce is a liquid condiment in which fermented soybeans are the primary ingredient. Homemade soy sauce is healthier and tastier than the store-bought stuff. However, it takes at least six months to make and it stinks while doing so.

There are two primary types of soy sauce. Shoyu is soybeans and wheat. It is "normal" soy sauce. The other type is tamari, which is a byproduct of making miso and is wheat free.

These instructions will make you 4L (1 gal) of shoyu. If you want tamari, make miso and scoop up the liquid that pools on the top of it.

What You Need

- 4 cups of dried yellow soybeans
- 4 cups of wheat flour
- Koji starter
- 3.5 cups of salt
- 4L of filtered water
- A pot
- A blender
- An oven tray
- Cheesecloth
- A large glass jar with lid
- A wooden spoon

Directions

The first step is to make koji. Soak the soybeans for at least 12 hours, then drain them and bring them to a boil. Lower the heat to a simmer and leave the beans to cook for about five hours, until they are very soft. Let them cool to room temperature.

Blend the cool, cooked soy beans into to a paste, then mix them with the flour until a dough forms. Add the koji starter and mix it well.

Spread the soy dough about 5cm (2in) thick on a tray. Make sure it is spread evenly. Make furrows in the dough with your finger, about 5cm (2in) apart and almost to the bottom of the tray. Cover the tray with cheesecloth and put it somewhere warm and humid for two days, such as on top of your fridge.

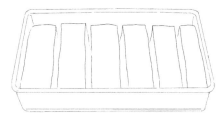

After two days, the koji will be ready. Now you will ferment it.

Mix the salt and filtered water together and stir it in well. This is your brine.

Put the koji in the large glass jar and fill it with your brine. Stir the koji around, but do not break the blocks up. Cover it and let it sit. Stir it once a day for a week, then once a week for at least six months. The longer you leave it to ferment, the deeper its taste will be.

When you are happy with it, strain the liquid through your cheesecloth into a pot. Squeeze as much liquid as you can out of the koji. When you're done, you can throw the solid koji away.

Heat the liquid to 80C (176F) and keep the temperature steady for 20 minutes. This pasteurizes the soy sauce to kill off any pathogens and gives it a longer shelf life. Allow it to cool and store it in sealed bottles in a dark and cool place. It will last up to three years.

Related Chapters:

- Miso

DOENJANG AND GUK-GANJANG

Doenjang is the Korean version of miso (with a stronger flavor) and guk-ganjang is the Korean version of soy sauce. The great thing about making doenjang and guk-ganjang is that it happens at the same time. That is, guk-ganjang is a byproduct of doenjang. Good doenjang takes at least one year to make.

What You Need

- 2.2kg (5lbs) of dry, yellow soybeans
- 8 cups filtered water
- 2 cups high-quality salt
- A blender
- A pot
- An electric heating mat
- Twine
- Chilies
- Goji berries
- Korean wood charcoal
- Large glass jars or a large Korean earthen pot (an onggi)
- Cheesecloth

Directions

It is important to wash your hands well before handling the doenjang or guk-ganjang at any stage of this process.

Soak the soybeans for at least 12 hours, then drain them and bring them to a boil. Lower the heat to a simmer and leave them to cook for about five hours, until they are very soft.

Mash or blend them to a paste while they are hot. It's okay if some are not ground.

Once they're mashed, make one large block then divide it into three blocks of equal size. You need to dry these blocks out. Put them on the electric heating mat, on the medium heat setting, to speed the drying process up. Turn them every three hours or whenever you can. You don't have to turn them while you sleep. White bacteria will form on them. That is good.

After the blocks have been on the heating mat for four days, hang them up. Make sure they are not touching. You can dry out chilies, which you will use for flavoring later, at the same time, but that is optional.

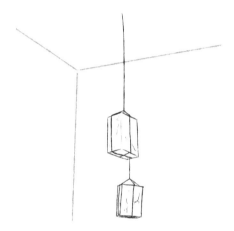

After the blocks have been hanging for six weeks, take them down. Put a layer of hay on the bottom of a cardboard box and place the blocks of doenjang on top of the hay. The hay will attract bacteria. Make sure the doenjang blocks do not touch each other. Close the box and place the electric heating mat on top of it. Turn the mat to low heat.

Leave the blocks in the box for two weeks, then hang them again for one more month. By this stage, they will be light but hard, with many fungus colors. They will smell.

Wash the fungus off quickly by scrubbing the blocks under running water. Shake the water off, then put them in a ventilated and sunny spot indoors to dry. Leave them for 24 days.

Make brine by mixing the filtered water with the salt. Stir the salt in to dissolve it. An egg should float in it. If it doesn't, add more salt.

Sterilize the glass jars or onggi with boiling water then place the doenjang blocks in them. Fill them up with the brine. Add a few goji berries, dried chilies, and Korean wood charcoal blocks. Cover the jars with a cheesecloth (tie the cloth in place with twine), and the lid and put them in a sunny spot, either in the garden or next to a sunny window. On sunny days, take the lid off. Do not let any water get inside.

In a few months, the blocks will turn an amber color, expand, and grow fungi. In another two months, the fungi will float. At that stage, take out the blocks and put them aside. Take out the chili, charcoal, etc., too. You can discard those. Strain the liquid through cheesecloth into a clean jar. Clean the tops of the original jars, but don't worry about cleaning the bottoms of them.

Knead the doenjang like dough until they are smooth and soft, then return them to their jars. Add one cup of the liquid to the jars. Sprinkle a little salt on top and cover them with the cheesecloth and their lids. Wait at least five more months. At that point, the doenjang will be ready. It does not need refrigeration.

With the rest of the liquid, you will make the guk-ganjang. Boil the liquid for 10 minutes. This will smell a lot. Strain it and return it to the jars. Allow it to ferment for at least five more months. At that point, it will be ready. If you leave it for a long time, some fungi will float on top. Just remove that. Keep the jars sealed and it will never go bad.

Here is the video I got most of this information from:

https://www.youtube.com/watch?v=PGhKwCq7SZk

If you are going to make this, I suggest watching it. She is quite entertaining, and you will see exactly how to do it.

Related Chapters:

- Miso
- Soy Sauce
- Charcoal

CHICKENS

Raising chickens is the easiest way to keep any type of livestock. They do not need much room (the average size yard is enough) and are relatively easy to care for.

In exchange for keeping them safe and fed, you will get an ongoing supply of fresh eggs and, if you want, fresh meat too. As a bonus, they are quite entertaining.

RAISING CHICKENS

Raising chickens isn't hard, but you have to get them past the chick stage first. That isn't too difficult either.

When they're chicks, the main thing is to make sure that they get food, water, warmth, a clean living environment, and some room to move around. While they are young, this takes a bit of monitoring, because they will walk and poop everywhere, including in their food and water.

Once grown, they will still need the same things, but will be more self-sufficient.

This chapter will give you an overview of how to raise chickens from chicks to adults. It is general information. Some breeds have different characteristics and needs. For example, some are happier in coops and some prefer to be free-range. Research the best chickens to keep for the environment you live in and what you want (eggs, meat, and/or brooding).

When you first start out raising chickens, it is best to keep your operation small. Two to three chicks are a good starting point. Before you buy your chicks, get everything you need to raise them. For hygienic purposes, always wash your hands after handling chickens, eggs, their home, and anything else to do with them.

Brooder

A brooder is where your chicks live until they are old enough to go into the coop. Keep it in your home so they are protected from the elements.

You can make it out of any large box. A cardboard box, a plastic storage container, or a cage will all work. It must be big enough to hold their food, water, and have room left over for them to move. Use a screen for the lid to prevent them escaping while keeping adequate air circulation. It will also protect them from pets and anything falling inside.

Layer the bottom of the brooder with clean pine shavings. This is the chick-litter. You will need to change this at least every couple of days. Never let it remain damp for more than 24 hours.

Get a medium-sized waterer and place it in the brooder. A plastic one is cheap and easy to clean. Raise it off the ground a little to prevent poop getting in, but not too much—they need to be able to drink out of it. Clean the waterer and change the water at least once a day.

You also need a feeder. Use medicated chick feed (called crumbles) to prevent diseases. After 10 days you can give them extra treats, like worms or bugs. Clean and refill the feeder often.

Keep the brooder warm with a 100w light bulb. Make sure the chicks can't get too close to it, as it may burn them. For their first week, keep the temperature around 35C (95F). You can use a small electronic temperature gauge to check. Reduce it by a few degrees each week until they have feathers, which will be in about six weeks. To do that, move the light away a little, use a dimmer switch, or change out the bulb to a weaker wattage. Do the opposite if they are too cold.

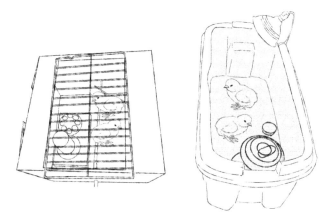

Example brooders.

If they huddle away from the light, they are too hot. If they huddle together under the light, they are too cold.

At the one-month stage, set up a low roost in the brooder so they can jump and/or sleep on it. At that stage, if the weather is warm, you can take them outside to play. Be careful of any dangers such as pets, snakes, or small holes they might fall into.

Pasting

Pasting is a condition where a chick's loose droppings will clog up its bum. You need to check for it often. When you come across it, clean up the droppings with a moist towel.

Chicken Coops

Once your chicks have feathers, you can move them outside to the coop. Make sure to move them during the warm season.

If you already have chickens, introduce the new ones slowly. Do it for an hour or two on the first day, and then gradually increase the time they spend together.

You can buy or make your chicken coop. Coops come in all shapes and sizes (search the internet for inspiration). No matter what design you choose, it has to provide your chickens with safety from predators, protection from the elements, and enough room for them to move around. As a rule, they need at least 1m (3ft) square per chicken. If you're using a chicken tunnel, have at least 3m (10ft) per chicken.

You may also need to add sun lamps for warmth and light, depending on where you live. Likewise, make sure there are some shaded areas for them to go in hot weather.

Always make sure they have access to water. If there is a chance of their water supply freezing, invest in a heated waterer or find some other way to keep it thawed.

There is specific feed for egg-laying hens. You can also give them leftover vegetable scraps and let them eat whatever bugs and weeds they find. Never feed them junk food or let them eat their eggs.

Nesting Boxes

A nesting box is a place for your hens to lay their eggs. You need at least one box for every four hens.

A crate laid on its side (with the opening on the side) makes a good nesting box, but you can use any similar box that is at least 30cm (10in) square. Place it in a dark and cool (but not cold) place, 50cm (20in) off the ground and at least 1m (3ft) from the roost.

Make it comfortable for the hens with pine shavings for warm climates and straw in cold climates. A layer of bedding 10cm (4in) thick is good. Clean out the bedding weekly. You can put it directly in the compost.

If an egg cracks, you need to clean the bedding out as soon as possible. You do not want your hens getting a taste for their eggs; if they do, they will start to peck at them.

Playtime

Bored chickens will peck their eggs, so make sure they have enough space to move around and things to do. Some ideas are:

- Hang a cabbage for them to peck at.
- Make obstacles for them to jump on such as a tree stump, ladder, and/or swing.
- Give them a pile of hay; they will keep busy evening it out.

Related Chapters:

- Chicken Tunnel

CHICKENS FOR EGGS

To get the best quality and quantity of eggs from your chickens, give them:

- Nutritious food
- Free-range time
- A stress-free life (security, warmth, etc.)

You do not need to keep a rooster for the hens to lay eggs. A rooster is only needed if you want to breed chickens. If you do have a rooster but don't want chicks, be sure to collect and store the eggs frequently.

Depending on the breed, hens will start to lay once they are around six months old. When there is enough daylight, expect one egg per day, per hen. When the weather is cold or when the hens malt, you will get fewer eggs. You can add lights to simulate summer, but it is better to let nature take its course.

Collection

Collect the eggs at least two times a day, and more often if it is hot. This will prevent the hens from pecking at them and/or getting a taste for any cracked ones.

Discard any cracked or rotten eggs. If they float, they're rotten. Clean the rest as soon as possible. Scrub them with warm water and unscented dishwashing liquid. Even if they look clean, you need to wash them. Dry them, place them in a container for storage, and date the container. Eggs will stay good for more than three weeks in a fridge.

When you have an abundance of eggs and you don't want to sell or give them away, you can freeze them. Crack them out of their shells into an airtight container, label them with the date, and freeze them.

Weak/Thin Shells

Weak/thin eggshells indicate a calcium deficiency. Supplement your chickens' food with store-bought calcium. Fresh whole milk and/or ground oyster shells are fine, but do not give them egg shells.

Laying Yolk

Laying yolk only is normal when your hens are first starting to lay, and it may take a couple of weeks before you get actual eggs. Make sure you clean the yolk out.

Egg Pecking

If a hen starts eating or pecking at eggs, isolate her before she teaches the others. You can identify which one it is by looking for yolk on their beaks. One way to train this habit out is to create a mustard egg. Make a hole in the top and the bottom of an egg and blow out all the raw egg. Fill it back up with mustard and put it in the nest. Keep doing this until the hen(s) learn(s) that eggs taste bad.

When you are unable to train a hen out of the habit, you need to keep her isolated or get rid of her.

Lost Eggs

Sometimes hens like to lay eggs in strange places, especially if you let them range freely before midday. You need to watch where they like to lay them. Watch them early in the morning. When you hear loud cackling, and/or when a hen seems restless or cranky, it is a good indication that she will lay soon.

Most often, a hen will lay in a "safe" place that offers coverage for the egg. Bushes, logs, the bases of trees, in long grass, in buildings, and along fence lines are all prime places.

Once you find the rogue eggs, put one in each nesting box so the hens see where to lay. Mark the eggs so you know which ones not to collect.

Egg-Bound

A hen is egg-bound when she is unable to pass an egg because it is stuck in her oviduct. This may kill her. A good diet with a safe environment will usually prevent this, but it can also be hereditary. Any of the following signs is an indication a hen is egg-bound:

- The hen has constipation or diarrhea.
- She goes in and out of her nest repeatedly.
- She isn't eating/drinking.
- She's not laying (though she might just be taking a break because of the heat, for example).
- Her comb and wattles are pale.
- She's exhibiting strange behavior.
- Her tail is pumping up and down.
- She's waddling.

Many of these signs can indicate any illness in a hen. The last two signs (waddling and tail pumping) are strong indications she's egg bound, but they can also mean the chicken is constipated.

To know for sure, you need to look or feel for the egg. You may be able to see it. If not, put on a latex glove and coat one finger with KY jelly. Gently insert this finger into the chicken's vent (bum/egg hole/cloaca) then push it straight back about 5cm (2in). If you can feel the egg, she's egg-bound.

Once you identify an egg-bound hen, you need to help her lay the egg. You need warm water, Epsom salts, and either vegetable oil or petroleum jelly. Handle her gently so she doesn't get stressed out. If she doesn't like something, stop doing it.

Give her water with electrolytes during the treatment period.

Sit her in a warm tub of water or a steam room for half an hour. You can make a steam room in your bathroom by turning the hot water in the shower on. 30C (85F) is a good temperature for the room or the water.

Rub her abdomen gently. Do so lightly so that the egg doesn't break. It is very important to avoid breaking the egg inside her.

Rub the vegetable oil or petroleum jelly on her bum. Wear latex gloves while doing it and then leave her alone some place warm (like the steam room) for 30 minutes.

When none of the above things work, or if yolk comes out, call a vet. Your vet may advise you (or you may choose) to cull her. It is either that or she will live in pain.

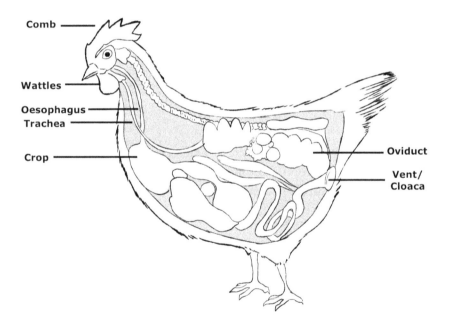

Stops Laying

As hens get older, they will start to lay less. Expect five to seven years of productivity, depending on the breed. Sometimes young hens will stop laying too. Some signs of hens getting to that stage are:

- Excellent feathers (their body can put more resources into them since they aren't laying).
- More pigment in the legs.
- Pale comb and wattles.

Even when a hen stops laying, she is still useful. She will always continue to create manure and eat bugs and weeds from your garden. Some hens may be good for brooding and may even produce eggs occasionally.

If you don't want to keep a hen, you can either give her away, release her into the wild, or cull her for meat. Older hens have tough meat, but it is good for stews and soups. Make sure you cull her humanely as explained in the *Culling Chickens* chapter.

Related Chapters:

- Culling Chickens

AUTOMATIC WATERING TUBE

Normally you need to clean and refill your chickens' water supply every day. Creating an automatic watering tube will eliminate this job.

With this design, your chickens can peck at the water bucket to get a drink. The lid will keep the water free from contaminants and the need to refill it will be much less than normal. You can also make it auto-refilling if you want.

What You Need

- 20L (5 gal) bucket with a lid
- Horizontal chicken nipples. Vertical nipples placed horizontally will leak.
- Drill and drill bit. The size depends on how big the nipples are.

Directions

Drill holes around the bucket a couple of cm (1in) up from the bottom. You need one hole for each nipple you want to add. Space them evenly around the bucket. Attach the nipples according to the manufacturer's instructions to prevent leaking and damage.

Place the bucket somewhere it is easy to fill up. A 20L bucket of water weighs 20kg, so you don't want to have to lug it too far. Sit it on something at a good height for the chickens, like a few cinder blocks, and fill it with water. Keep the lid on it to prevent contaminants from getting in.

To make it refill automatically, connect it to your water supply and use a stop valve.

Take it apart and clean it once a month and at any time it starts to leak.

Making something similar for chicks is possible, but not recommended, because to do it you need to use vertical nipples. The chicks are not strong enough to push the horizontal design. This means that you would need to train them to use the vertical nipples, and then retrain them to use the horizontal ones. If you want to do it anyway, use a much smaller container and install the vertical nipple on the bottom of it.

Training Your Chickens to Use a Water Nipple

When switching your chickens to an automatic waterer, do it during cool weather. The younger ones will learn quicker, but older ones may take some time, especially if they can't see well.

Remove all other water sources and demonstrate to them how it works. Once one figures it out, the rest will learn from it.

CHICKEN TUNNEL

Chickens that have space to run around in are happier, and a happy chicken will produce better eggs.

A chicken tunnel allows them the space to run around without giving them free rein of your garden. This is good if you have a small coop and/or need to protect them, or protect something from them.

Chicken tunnels are also easy to move, so you can put yours wherever you want your chickens to go at different times. For example, if there is land you want to prepare for gardening, they can eat the weeds and bugs while fertilizing the ground.

Example of a chicken tunnel.

What You Need

- Chicken wire.
- Fencing stakes.
- Wire supporters. These keep the wire in shape.
- Wire cutters.

Directions

First, plan out where you want to put the tunnel. It must start at the coop and either loop back into itself or get closed off at the end. The entrance must be bigger than the coop door.

Once you have your plan, measure how much wire and how many stakes and supporters you will need, and then go out and get them.

Roll out the chicken wire where you want to place the tunnel, and secure it to the ground with the stakes. Add the supporters to shape the wire and attach it to the coop opening. You can use a wooden or plastic panel to block off access at the coop entrance when needed.

BREEDING CHICKENS

At some stage, you may be interested in breeding chickens. This is a good way to sustain or expand your flock, make some extra cash by selling them, and/or to produce chickens for their meat.

Approximately 50% of chicks will be male. Besides having the disadvantage of not producing eggs, males eat more and are noisier. They also like to fight each other, so you need to keep them separated. Unless you want to eat or sell chickens, you are better off buying hens from a breeder or adopting rescue hens as opposed to breeding your own.

If you do want to breed your own, aim for one rooster for every 5 to 10 hens.

When choosing a rooster, go for a non-aggressive breed. It does not matter if your hens are not the same breed as the rooster, unless you want a specific breed of chicks. Look for a rooster that is healthy overall with no deformities and an even eye color. Give them good-quality food and then let nature take its course.

Once you have a rooster, choose the eggs you want to hatch. Pick the best ones and handle them extra carefully. Look for the following characteristics:

- Average size
- No cracks
- No thin shells
- Regular shape

Wipe any poop and mucus off them, but do not clean them like normal, as that will remove the protective film. You can mark them to keep track of the breed and dates, but this is not necessary for the average household coop. Expect 70% of them to hatch.

There are two ways to hatch chicks. You can let a hen hatch them or do it in an incubator.

Hen Hatching

The most natural way to get chicks is to let a hen hatch and raise them. To do this, you need a "broody" hen that will sit on the eggs. You also need to be prepared not to get any eggs from that hen for about three months.

Some hen breeds, such as the Silkie and Cochin, are broodier than others, but any hen may be broody. Some signs of a broody hen are:

- She's defensive.
- She's constantly sitting on the eggs.
- She walks with her feathers raised and her head low, and clucks frantically.

To test if a hen is broody (or to train her to be broody), sneak a fake egg or two underneath her while she sleeps. Golf balls make good fake eggs. If she sits on them for 24+ hours, she will probably go the long haul.

If she doesn't sit on them but you want her to be broody, keep the fake eggs in her nest until she starts sitting on them, which may or may not happen. If/when it does, swap the fake eggs out for real ones. Another method is to just let her lay a few eggs. She may naturally become broody when she has three to five eggs in her nest.

Once you find a broody hen, separate her (with her nest) from the other hens by moving her to a brooding pen. Do it at night if you need to. Create a peaceful environment for her. Make it quiet and dim. Give her some time to get used to the new surroundings, then sneak eggs under her. Depending on her size, she can hatch 6 to 12 eggs at a time. She must completely cover all the eggs. If not, take some away.

Make sure she eats while she is hatching the eggs. You may need to put food and water right next to her. You can also switch her to chick feed.

Hatching will start in around three weeks. Do not disturb her while the eggs are hatching. Remove any eggs that are still unhatched after 72 hours of the first one hatching. Once the chicks have hatched, the hen will raise them. Keep them separated in the brooding pen for six weeks.

Usually, keeping multiple brooding hens together in the same brooding pen is no problem, especially if they're introduced at different times. Separate them if they fight.

Incubating

To incubate the eggs yourself, you first need to store them, pointy end down, at 13C (55F) for a minimum of 24 hours and a maximum of 6 days before trying to incubate them. The higher the humidity, the better. Turn them once a day to prevent the membranes from sticking. You can do this without touching them by raising one side of the carton you store them in and then switching the raised side each day.

Put the incubator in a temperature-controlled room and operate it as per the manufacturer's instructions. Alternatively, see the next chapter to make your own.

Don't help chicks that can't hatch. It is survival of the fittest. If they are not strong enough to hatch, they will probably die anyway. Once a chick has hatched, wait until it is dry and at least one hour old, then move it to your brooder.

Are the Chicks Male or Female?

You need to wait at least two months to figure out of a chick is male or female. Don't try to look at their genitals. You will damage them.

You can tell a male because his comb and wattles are larger and redder, and his legs are chunkier than the females'. He may start to crow at 12 weeks and will strut with a puffy chest and his head in the air.

Some breeds can be sexed earlier. Research your breed to find out their telltale signs.

Roosters

Roosters are fine together when they're young, but they will fight when they're older. You need to either separate them or get rid of all the ones you don't want by giving them away, selling them, or culling and eating them.

Roosters need perches, and they will crow. There are ways to prevent roosters from crowing, but all of them are cruel. If you don't want the noise, don't breed or house roosters.

Inbreeding

Once you have your first batch of chicks, you need to be careful of inbreeding. To control this, you can either separate the rooster from its offspring or replace the rooster.

STYROFOAM INCUBATOR

This Styrofoam incubator is not as effective as a store-bought one, but it is easy and cheap to make, and will do the job. It is a good option for your first try at incubating. If you want to continue doing it, you can splash out on a proper incubator.

What You Need

- A Styrofoam box.
- A Light bulb socket that plugs into a standard extension cord.
- An incandescent light bulb. The wattage will depend on the size of box. 25w is good for a medium-sized box.
- Chicken wire.
- A hygrometer. This is a thermometer and humidity gauge in one. Get a good-quality one.
- A small bowl of water.
- A sponge.
- Pebbles.
- Glass from a picture frame. Get a piece that is slightly smaller than the box lid.
- A screen, hardware cloth, or fabric to wrap over the frame.
- Masking tape.
- A pencil.
- Fertilized eggs.
- A flashlight.
- A brooder.

Directions

Cut a hole near the top of the Styrofoam box for the light socket. Make it as snug as possible. Insert the socket and tape around it on both sides as a fire precaution.

Put a layer of clean pebbles on the bottom of the box. This helps with temperature control.

Use the chicken wire to make a wall between the bulb and the rest of the box. This is so the chicks won't burn themselves once they hatch.

Put your hygrometer on the side where the eggs will be, and put the small bowl of water with the sponge on the opposite side.

Cut a hole in the box lid a little smaller than the glass, and tape the glass over the hole as a viewing window.

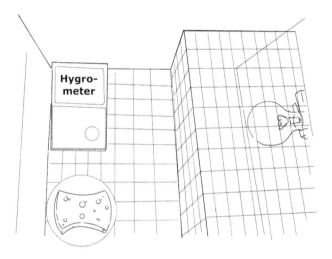

Turn on the light and monitor it for 24 hours. You want a temperature of 37.5C (99.5F) and 40 to 50 % humidity. Adjust the heat and water until you reach those levels. Some ways to do this are:

- Adding/removing water in the sponge for humidity.
- Installing a mini fan to keep the temperature more consistent.
- Making holes in the side of box to cool it. Tape them up if it's too cold.
- Use a dimmer switch on the light help regulate temperature.

- Wiring a water-heater thermostat to the power source so the lightbulb will automatically turn off/on at particular temperatures.

When you are ready, put the eggs in it. Always handle the eggs carefully and with clean hands. Mark one side of each egg with a "1" and the other with a "2." Put them in so "1" is facing up. Group them together to help keep a constant temperature.

Monitor the humidity and temperature three times a day, and at the same time, turn the eggs over. Make sure to alternate which side is up each night—that is, "1" the first night, "2" the second, "1" the third, and so on. If you are handy, you can make a mechanical rotator for the eggs.

After seven days, check if the eggs are fertile or not. Hold a flashlight up to each egg in a dark room. Fertile eggs will have a dark spot with blood vessels. An infertile egg will have a ring or streak of blood, and/or light up bright and evenly. Put the fertile eggs back in the incubator and throw away the infertile ones. If you are unsure about an egg, put it back.

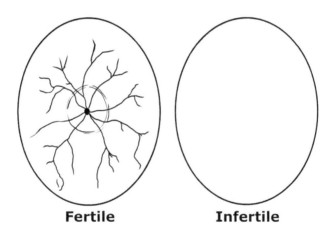

On the 18th day, increase the humidity to 65% to 75%.

Hatching will start around three weeks. Do not disturb the eggs while they are hatching. Remove any eggs that are still unhatched after 72 hours of the first one hatching. Don't help chicks that can't hatch. It is survival of the fittest. If they are not strong enough to hatch, they will probably die anyway.

Once a chick has hatched, wait until it is dry and at least one hour old, then move it to your brooder.

CHICKEN BATHING

Under normal circumstances, a chicken will keep itself clean. If it is injured, ill, or dirty (if it has dry poop on it, for example), then you may want to bathe it.

What You Need

- 2 large plastic or metal tubs.
- Unscented "sensitive skin" dish soap.
- Warm water. Hose or tap water is fine if the ambient temperature is not cold.
- Salt (optional).

Directions

Clean both of the tubs.

Fill one of the tubs with warm water and add five drops of the dish soap. Make sure it is not too soapy. If the chicken has mites, also add a few tablespoons of salt. Fill the other tub with warm water.

Grab the chicken so you are holding its wings down and place it in the soapy water. Be gentle and keep its head out of the water. Lather it up. If there is a dried stain and/or if the chicken has mites, soak it for five minutes. You can remove any grime from the chicken's nails while you're waiting, but that it optional.

Once it is clean, transfer the chicken to the water-only tub. Rinse off all the soap with a cup and your hand.

Wrap the chicken in a towel and/or gently dab it dry. Using a blow dryer on the cool setting will dry out under the feathers faster. Keep it at least 15cm (6in) from the chicken to avoid burning it. Once the chicken is completely dry, release it back into the coop.

CULLING CHICKENS

At some stage, you will probably need to cull a chicken, whether it is ill, injured, or you want eat it. If you are culling it because it is ill, don't eat it. Make sure the other chickens can't access the area where you're doing the culling, or they'll come and grab all the leftovers.

Hypnotizing the Chicken

You do not have to hypnotize a chicken before culling it, but it's easy to do and will make your job easier.

Lay the chicken on its side so its wing is pinned under its body. Using your finger, tap once right in front of its beak and then again 15cm (6in) in front of its beak. Alternate this tapping until the bird stays still.

Alternatively, hold the chicken down on its breast and gently press on its back. Move its legs back if it tries to stand. Draw a line on the ground with a stick. Start from right in front of its beak and draw it back about 30cm (1ft).

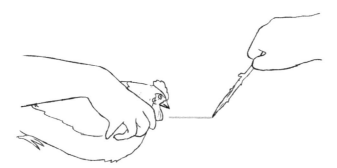

Once the bird is hypnotized, kill it as soon as possible.

If you are hypnotizing it for any other reason, expect it to wake up within a few minutes. Otherwise, you can wake it up by clapping.

Culling

To cull the chicken, you need:

- A sharp axe
- A Stump
- 2 nails
- A hammer
- A bucket
- Old clothing

Hammer the two nails into the stump, just far enough apart for a chicken's neck to fit. This will hold its head in place.

Hypnotize the chicken and put its neck between the nails. Pull on its legs to stretch out its neck, then cut its head off with the axe. It will flap, but it will be dead.

Hold it upside down over the bucket to drain its blood. This will take a few minutes. Alternatively, you can get a cone and put a bucket underneath it, then slit the chicken's throat while holding it over the cone.

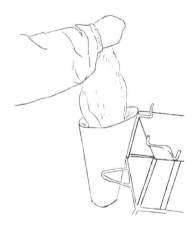

Plucking

Remove the chicken's feathers immediately after culling it. You will need:

- A bucket of scalding hot water
- A bucket of ice water (optional)
- Heavy-duty rubber gloves
- A butter knife

Once all the blood is drained, dunk the chicken head-first into scolding water and gently move it around and up and down. Soak all its feathers, including those on its legs.

After 30 seconds, remove the chicken from the water and rub your gloved hand against the grain of the feathers. If they come out easily, it is done. If not, soak it for another 30 seconds and recheck. Doing this in 30-second intervals is important, because if you soak it for too long, it will damage the chicken's skin.

Once it's ready, wipe the feathers off by rubbing your hands against them. Pluck larger wing and tail feathers one at a time to avoid tearing the skin.

Use a dull butter knife to scrape off the remainder. Special pinning knives are available if you want.

Rinse the plucked chicken.

If you're culling more than one, place them in ice water until they are all ready to process.

Skinning (Alternative to Plucking)

If you don't want the skin of the chicken anyway, you may as well skin it as opposed to plucking it. You will need:

- Somewhere to hang it
- Twine

- A sharp knife

After killing and draining the chicken, tie its ankles together and hang it at head height, with its breast facing you.

When doing this, you want to make cuts just deep enough to separate skin, but without cutting into the flesh. Cut the skin around the chicken's leg joint, but be careful not to cut the tendon. Slice down the front of the leg as you pull the skin down. Do the same with the other leg.

Slice from leg to leg across the bottom of the abdomen. Pull and cut down the front to the neck. Do the same for the wings, but cut them off at the first joint. They hold little meat, so it is not worth saving them. Remove the feathers and skin in one piece.

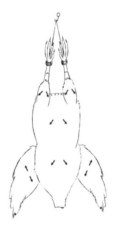

Gutting

Once it is plucked or skinned, you can gut the chicken. You will need:

- A large chopping board
- A sharp knife
- A bucket of water

Chop off the chicken's feet between the joints. These are great for broth.

Cut out the preen gland. If you keep it in, it will ruin the taste of the meat. Cut in above the gland down to the bone, then slit along the bone to the tail. Ensure there is no yellow glandular tissue left on the bird.

Slit the skin along back of neck (if it has skin) and slide the skin down. Separate the trachea and the esophagus from the neck, then pull the crop out of the body. Leave the trachea, esophagus, and crop hanging out.

Insert the knife 2.5cm (1in) above the vent (bum/egg hole) and slit up to the breastbone. Be very careful not to cut internal organs.

Cut around the vent, then scoop the innards out with your hand. Put them into a bucket with water. Remove the trachea, esophagus, and crop if they didn't come with the innards.

Make sure you get all the lung tissue which may break. It is squishy.

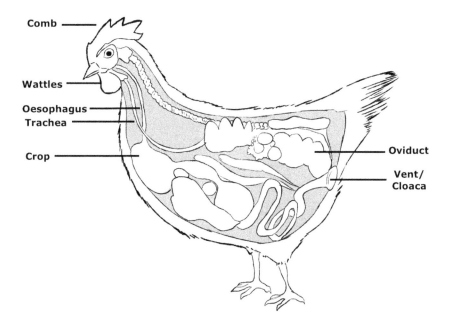

Stripping (Alternative to Gutting)

Stripping a chicken will give you pieces of meat like those packaged in the supermarket. You will not have to gut the chicken, but you'll waste some of the meat. You will need:

- A large chopping board
- A sharp knife

Cut off the chicken's wings as close to the breast as you can. Cut through and around the joint. Slice down the ridge between the breasts. Fillet as close to the breast bone as you can on both sides. Push the thigh and drumstick back so the thigh joint pops free of the hip. Cut off the leg and thigh as close to the body as possible. Cut off the back upper thigh about 5cm (2in) from the vent. Chop off the lower legs at the ankles.

Storage

Unless you plan to boil the chicken immediately after culling it, place it in ice water for an hour. Pat it dry, then put it in a container and refrigerate it for 18 hours before cooking. This will relax its muscles and make the meat tender. To store it long-term, freeze it after the ice bath.

Cleanup

Use the bones for broth or throw them away. Do not compost them.

If the chicken was not ill, you can feed some of the leftovers to your pets. Dogs, cats, and chickens can eat everything but the bones, intestines, and feathers. Cook anything you want to give them first, to be safe. Large amounts of organ meat may cause diarrhea, so if you are culling a lot, don't give it all to your pets.

You can compost the blood, guts, and feathers. Mix them with three times the volume of dry wood chips, chuck it in the compost, and cover them with sawdust or dry leaves.

When you do not have a compost heap, or if you have scavengers (foxes, raccoons, dogs, etc.) that will get to your compost, bury the remains in a 20cm (8in) trench. Spread them on the bottom of the trench.

The final way to get rid of the remains is to double-wrap them all (except the blood) in plastic (minus the blood) and throw them away in your garbage. Pour the blood straight into the dirt.

Clean everything you used with a diluted bleach solution, then rinse off the bleach.

BEE KEEPING

Keeping bees will give you an endless supply of the purest honey you can get, which is a delicious and versatile product with an indefinite shelf life. Honey is a healthy form of sugar. You can also use it as an antibacterial to protect skin, help heal minor wounds and burns, and boost your immune system when you are sick (you can put it in tea, for example).

Keeping bees will also provide you with beeswax, pollen, propolis, and royal jelly, all of which are useful in their own way.

Finally, bees will help pollinate your garden, making it more abundant.

So unless people in your household or your neighbors are allergic to bees, starting a beehive is worth looking in to. Check to make sure it is legal to do so where you live before you start.

STARTING YOUR BEE HIVE

Start your beehive in the spring.

What You Need

- Protective clothing, including gloves, veil, and jacket. Alternatively, you can opt for a full beekeeper's suit.
- A hive tool. A flathead screwdriver works, but a proper one is better.
- A spray bottle filled with syrup (one part sugar and two parts water). Make sure bottle is new, with no chemical residue.
- A top feeder.
- A queen catcher.
- A queen muff to keep the queen from flying away.
- Bottom board for the hive to go on.
- The hive. You can buy it new or secondhand, but get a good one. Buying a new hive kit is recommended for beginners.

Directions

Choose where to put your hive. Look for somewhere:

- Away from pets and neighbors.
- Dry.
- Near a source of standing water where the bees can stand while drinking, like a birdbath.
- Near flowering plants, preferably purple ones.
- Next to a high wall.
- Sunny.

Face the entrance towards the wall. That will force the bees to fly up when they leave the hive.

Once you have all your equipment, buy your bees. Don't do it beforehand, because if your bees arrive and you're not ready, it will be a problem. Get a gentle breed, like Buckfast or Italian.

Buying a nuc or two is easier than getting packaged hives, since the bees will have already accepted the queen. Ask to have the queen marked. Make sure you order your bees from a reputable seller that is not too far from you to cut down on their travel time.

The nuc will come in frames in a screen box. The best way to learn how to install your bees is to watch a video. There are plenty online. Here is one I found (assuming it is still live by the time you read this):

https://www.youtube.com/watch?v=70ODILM3qcM

In the video, he doesn't wear any protective equipment, but you should. Here's a rundown of the process.

First, you need to install the queen. Remove a few frames from your hive to make a gap where the bees can enter. Take off the plywood covering the entrance of the package using your hive tool.

Remove the feeder can and then the queen cage. Remove the cork on the candy side of the queen cage and attach the cage to the top of a frame in the hive. The candy faces down.

To install the rest of the bees, spray them with the sugar solution, then bump them down the package by tapping the package on the ground.

Remove the lid and spray inside the package. Place the package with the opening face-down on top of the hive where you removed the frames. Wait a few hours for most of the bees to leave the package, and shake out the rest.

Put the rest of the frames back in and close the hive up. Give the bees some food. You can also use an entrance reducer.

After one week, remove the queen cage, assuming the queen is out of it. If not, check back in a week.

Your bees can get diseases. Research your breed of bees and the area you are in so you know what to look out for and what action to take.

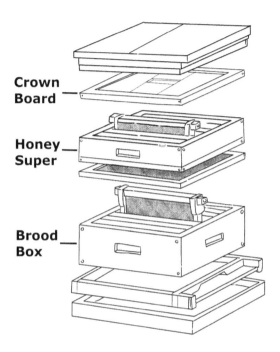

COLLECTING HONEY AND WAX

Honey and wax may be ready to collect at any time of year, but are most likely to be available in the fall.

Use the honey for eating, cooking, and health. Use the beeswax you can make things such as candles and salves.

What You Need

- Protective clothing, including gloves, veil, and jacket. You can also opt for a full beekeeper's suit.
- A smoker—the bigger, the better.
- A hive tool. A flathead screwdriver works, but a proper one is better.
- A bee brush or a feather.
- An extractor. This is for collecting the honey. It's expensive, so borrow or rent one to start.
- Cheesecloth.
- Jars, sterilized with boiling water.
- A bucket.
- A pot.
- Water.
- Containers.

Directions

Come up from behind the beehive and smoke the entrance, then open it and smoke the upper area. Remove the inner cove. If it is sealed with propolis, use your hive tool to pry it open.

Gently brush off the bees from the frame you want to take out, then extract the frame. Check to make sure most of the cells are capped. If not, the honey isn't ready and you should put it back. If it is ready, put it in an empty honey super while extracting the rest of the frames.

Once you have all the frames with honey, scrape the wax caps off both sides into a container. Put the frames into the extractor and rotate it. The honey will come out the bottom. Strain it through a cheesecloth and put it in sterilized jars.

Allow the honey to drain from the wax for at least two days, then put then drained wax in the bucket and top it up with warm water. Slosh the water around to remove any remaining honey. Strain out the wax and repeat this cleaning process until all the honey is off.

Put the wax in a sterile glass jar and place the glass jar in a small pot. Fill a quarter of the pot with water. Bring it to a simmer and stir while the wax melts. Once it is melted, remove the jar from the water, take out the wax, strain it through several layers of cheesecloth. Melt and train it again as needed until all the impurities are gone.

This jar-in-pot method is a makeshift double boiler. Double-boiling is used for safety when melting highly flammable substances.

Pour the clean wax into clean containers for storage. Beeswax gets very hard when it sets, so it is best to store it in small amounts. That way, you will not have to cut it later.

PRESERVING FOOD

When you produce more food than you can eat, you can either give it away, sell it, or preserve it for later. This section will cover the last of those options.

All the preservation methods in this book are natural, meaning there are no chemical preservative involved.

There are some risks with preserving food at home, because if you don't do it properly, microorganisms can grow, and these may make you sick. Do it at your own risk and follow the instructions carefully.

DRYING

Drying is the removal of excess water from food. It is easy to do, and you don't need any special equipment.

Dehydrating food is also the removal of excess water, but in that case, the heat and humidity are controlled. Dehydrated food lasts longer in storage than dried food, but it requires special equipment.

Whether you dry or dehydrate, the food becomes smaller and lighter. In most cases, when you want to eat the dried or dehydrated food, you need to reconstitute the water content, but there are some things you can eat without reconstituting them.

In this book, you will learn about drying food since it is more DIY than dehydrating. If you want to dehydrate your food instead, buy a dehydrator and follow the manufacturer's instructions.

There are several ways to dry food. Some are better for particular types of food. To ensure an even drying process:

- Cut the food into uniform pieces.
- Spray the drying racks with non-stick spray.
- Make sure the individual pieces don't touch each other on the drying tray.
- Turn them at the halfway point
- Check them more often near the end of the drying period.

Sun/Air Drying

Sun- or air-drying is only good for fruit, and it takes several days, longer in cool weather. Breezy days over 30C (86F) are the best. You also want minimal humidity. Anything over 60%, and it won't work well.

Place the food in the sun on any food-safe material and cover it with a cheesecloth to keep the bugs out. For the best results:

- Use a tray with all-round circulation (a mesh one, for instance).
- Put it over a reflective surface, such as tin or aluminum.
- Bring it in at night, so the night/early morning moisture doesn't slow down the process.

Here are a couple of examples:

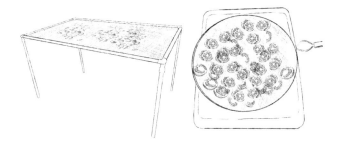

Solar

Solar drying is faster than sun/air drying and extracts more moisture. It is good for fruit and vegetables. Learn more about solar drying in the *Cardboard Box Solar Dehydrator* chapter.

Oven

An oven is the next best thing to a commercial dehydrator. With an oven you can dry fruit, vegetables, and meat. Set the oven to 60C (140F) and turn the fan on. If your oven is not fan-forced, leave the door open a little—5cm (2in) or so.

Pasteurization

Sun- or solar-dried foods need pasteurization before you can store them. You can either:

- Seal the dried food in freezer bags and freeze it for 48 hours; or

- Put it in single layers in the oven for 30 minutes at 70C (160F).

Storage

Once the food is pasteurized and cool, you can store it. Put it in a sealed container with some oxygen absorbers. Store the containers in a cool, dry, dark place.

- Dried fruits last one year.
- Fruit leather lasts one month.
- Dried vegetables last six months.
- The shelf life of dried meat depends on the type of meat and the method of drying.

All dried foods will last longer in the fridge or freezer.

Reconstitution

You do not need to reconstitute fruit or jerky, but you can if you want.

Cooking any dried food in soups and stews will make it edible. Soaking vegetables in boiling water will bring them back to a "fresh state." They may not look fresh, but as long as you don't over-soak, they will taste close enough to fresh, and you can use it as you would the fresh vegetable. You need three cups of boiling water for every cup of vegetables. Let them soak for up to an hour.

Simmer leafy vegetables and tomatoes in water until they're tender.

To reconstitute fruits, soak them in room -temperature water.

Related Chapters:

- Fruit Leather
- Cardboard Box Solar Dehydrator

DRYING FRUIT

When drying fruit, there several steps to go through:

- Preparation
- Pretreatment
- Drying
- Conditioning
- Pasteurization (if needed)
- Storage

Preparation

The minimum preparation for fruit is to wash it. Peeling is optional, but the drying process will take longer if you don't. Core fruits that need it, like apples, and slice them if you want. The smaller the slices, the faster the drying process.

When you don't want to slice or peel the fruit, you need to crack the skin. Put it in boiling water and then immediately in cold water.

Pretreatment

Pretreating fruit prevents it from darkening. There are quite a few ways to do it, but here are the easiest ones.

Make any of the following mixtures:

- 3000mg of pure ascorbic acid (Vitamin C) with half a liter of water.
- Half a cup of sugar with one and a half cups of boiling water. Once it cools to lukewarm, add half a cup of honey.
- 1 teaspoon (5g) of citric acid with one liter of water.
- Lemon juice and water in equal parts.

Soak the fruit in the mixture for 10 minutes then drain it.

If you plan on storing the dried fruit for an extended period, consider using a sulfite dip, though it can cause asthmatic reactions. Look for food-grade sodium bisulfite, sodium sulfite, or sodium meta-bisulfite in any wine-making supply shop. To make the mixture, mix either:

- 1 teaspoon of sodium bisulfite per 1L of water
- 3 teaspoons of sodium sulfite per 1L of water
- 2 Tablespoons of sodium meta-bisulfite per 1L of water

Soak the fruit slices in the mixture for five minutes, then rinse them under a light stream of cold water. You can only use your sulfite dip once. Make a new dip for each batch of fruit you will dry.

Drying

Dry the fruit using any method. Since the fruit won't be rehydrated, leave 20% moisture in it. You can test if it is dried enough by cutting a few cooled pieces in half and squeezing them. No moisture should come out.

Fruit slices should be pliable but not sticky. If you fold them in half, they should not stick to themselves. Berries should rattle when shaken.

Conditioning

Conditioning equalizes the moisture of dried fruit. This reduces the risk of mold growing during storage.

Pack the fruit loosely in a sealed jar and leave it for 10 days. Shake the jar every day. If you see condensation, you need to dry the fruit more. Once it is conditioned, pasteurize it if needed. Pack and store the fruit.

FRUIT LEATHER

Fruit leather is dried, pureed fruit. Two cups of fruit will make about a 33cm (13in) x 38cm (15in) rectangle of fruit leather.

What You Need

- Fruit
- A knife
- A chopping board
- Lemon juice or ascorbic acid (vitamin C)
- Honey (optional)
- Applesauce (optional)
- Greaseproof paper

Directions

Wash your fruit, peel it if needed, and cut it into chunks. Make sure there are no seeds or stems. Puree it with either two teaspoons of lemon juice or 375 mg of ascorbic acid. If you want to sweeten it, add one quarter of a cup of honey. Adding applesauce is optional. It will extend the puree, decrease the tartness, and make it easier to work with.

Line a tray with greaseproof paper, then spread the puree evenly on it so it is a couple of centimeters (1in) thick. Dry it any way you like. Here are some approximate timings:

- 7 hours in a commercial dehydrator
- 18 hours in an oven at 60C (140F)
- 48 hours in the sun

Test if your fruit leather is done by pressing the center of it with your finger. It should not leave an indent. Once it's ready, pasteurize it if needed. Pack and store it.

DRYING VEGETABLES

When drying vegetables, you want them to get brittle.

What You Need

- Vegetables
- A chopping board
- A knife
- A steamer or pot and a colander

Directions

Wash and cut the vegetables. Remove any bruised or spoiled parts.

Next, you need to blanch the vegetables. This prevents loss of flavor and color during the drying process. It also shortens the time needed to dry and rehydrate them.

Steam blanching is the preferred method as it keeps more nutrients than water blanching.

If you don't have a steamer, use a deep pot with a tightfitting lid and a colander. Fill the pot one quarter full with water and bring it to a rolling boil. Put the vegetables in the colander, maximum 5cm (2in) deep. Put the colander in the pot, but make sure the water isn't touching the vegetables. Cover the pot and let the vegetables steam.

Research the steaming times depending on the type of vegetable and the size of the pieces.

Once they're blanched, put the vegetables in cold water until they're cool enough that they are just a little too hot to touch. Drain the water and dry them using the method you want. When they're brittle, they're ready to store.

DRYING HERBS

Drying herbs is easy and will give them an indefinite shelf life—as long as you keep them from getting wet.

What You Need

- Herbs
- Twine
- A fan (optional)
- A paper bag (for tender herbs)
- Scissors

Directions

Pick the herbs from your garden in mid-morning, after the dew dries. Remove any damaged leaves and wash the good parts carefully. Gently shake off any excess water.

Make small bundles of the herbs and tie them together by their stems. Hang them in a ventilated area but not in direct sunlight. Put a gentle fan on the hanging herbs if you want them to dry faster.

Some "tender" herbs have more moisture. Common ones include basil, lemon balm, mint, oregano, and tarragon. Using a paper bag will help absorb the extra moisture.

Cut the top off the paper bag so it is a little shorter than the bunch of herbs. Use scissors or a hole punch to make a hole in the center of the bottom of it. This is the first hole. Make more holes all over the bag. Put the stems through the first hole so the leaves hang inside the bag.

When the leaves crumble between your fingers, they are done. This will take about five days if you don't use a fan, but it also depends on the type of herb.

Clean and dry your hands then remove the leaves. Place them in an airtight container and label it with the type of herb and date.

POWDERED EGGS

Powdered eggs are not as tasty as the real thing, so unless you have a big surplus of eggs or you want to take them hiking, it's probably not worth doing.

You can either dehydrate raw eggs or cook them first. Cooking them first is faster but tastes terrible, so these are instructions on how to powder raw eggs.

What You Need

- Eggs
- A bowl
- Greaseproof paper
- A tray
- An oven or a dehydrator
- A bowl
- A freezer
- A blender

Directions

Crack the eggs into a bowl and scramble them well.

Line a tray with greaseproof paper, pour the eggs on top, and then put them in your oven or dehydrator. Sun-drying is not safe due to the risk of salmonella.

The temperature of your oven or dehydrator needs to be 75C (165F). It is higher than normal to kill off salmonella. Not all dehydrators will reach this temperature.

The eggs are ready when they are completely dry and flaky. They will turn dark orange and have a greasy feel. Let them fall into a bowl and continue to dry anything that sticks to the tray.

Once the eggs are dry, pasteurize them in the freezer. Even though they were not sundried, you still need to pasteurize them to be safe.

After pasteurization, blend the eggs to powder. Store them in an airtight container with some oxygen absorbers.

Powdered eggs will keep for one month at room temperature or one year refrigerated.

To rehydrate the eggs, mix them one to one with water. Mix them thoroughly. One tablespoon of powdered eggs is equal to one egg, so if you want to eat the equivalent of one egg, mix together one tablespoon of powdered eggs with one tablespoon of water.

Let the mixture sit for five minutes, then cook it.

POTATO FLAKES

Making potato flakes is a great way to store extra potatoes. They are easy to cook, ultra-portable, and have a shelf life up to five years.

What You Need

- Potatoes
- A knife and a chopping board
- Water
- A pot
- A blender
- A tray
- Greaseproof paper

Directions

Peel, wash, and dice the potatoes, then boil them until they're cooked.

You need about half a cup of water for every large potato. Mix the cooked potatoes and water in the blender until they are thin and a little runny. Keep adding water bit by bit until the mixture is the right consistency.

Line a tray with greaseproof paper and spread the mixture evenly on top. It must be a maximum of 0.5cm (0.25in) thick. Dry the potatoes in an oven or dehydrator at 65C (145F) for seven hours or so. Check and turn as needed. If you are using an oven, make it a bit hotter and leave the door open a little to let the moisture escape.

When the potatoes are crunchy but not dark, they're done. Let them cool, then pulverize them in the blender. Store them somewhere cool, dry, and dark in an airtight container with some oxygen absorbers.

When you want to eat them, make mashed potatoes. Mix one part potato flakes with two parts water. Stir the mixture and let it sit for five minutes. Warm it up on the stove and add whatever you usually put with your mash.

CARDBOARD BOX SOLAR DEHYDRATOR

This is an easy and cheap way to make a solar dehydrator. If you decide you want something more permanent, you can make a better one with some basic carpentry skills (look for free designs on the internet) or buy one.

What You Need

- A cardboard box
- Matte black paint
- Food-grade mesh wire
- Duct tape
- A solar-powered mini-fan
- Bricks

Directions

Paint the cardboard box matte black inside and out. This helps to absorb heat. You can skip painting the outside on the bottom and one of the long sides.

Cut a large rectangle out of the unpainted long side and cover it with some mesh wire. Use the duct tape to hold it in place.

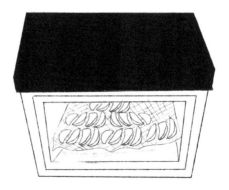

Make a hole on one of the small sides to put the solar-powered mini-fan in.

Place a few bricks inside the box and lay mesh wire on top of them. Make it so the mesh sits about half way up the box. This is the rack to place the food on.

Put a brick on top to hold the lid closed.

PICKLING

Pickling your excess produce can keep it good for years. The basic idea is to immerse the food in a vinegar brine, which is known as quick pickling. Unlike slow pickling (fermenting), quick pickling ensures that whatever you preserve is ready to eat on the same day you make it. Quick-pickled foods don't have the same health benefits as fermented foods, but you can make them with apple cider vinegar, which is itself a fermented food.

This recipe uses an even ratio of vinegar to water to make the brine. You can use more vinegar if you want, but not less.

What You Need

- 1/2kg (1lb) of fresh vegetables.
- 1 cup of vinegar with a minimum of 5% acidity.
- 1 cup of water.
- 1 tablespoon of rock salt.
- Herbs and spices, like garlic, peppercorns, bay leaves, or chilies, to taste(optional).
- Jars.
- A pot.
- A wooden spoon.

Directions

Clean and slice the fresh vegetables and pack them in jars with any herbs and spices you want to add. Pack them as tightly as you can without damaging them. Leave 1.5cm (0.5in) space at the tops of the jars.

Bring the vinegar, water, and salt to a boil. Stir the mixture to dissolve the salt. This is brine. Pour the brine over the vegetables, filling each jar to within 1.5cm (0.5 inch) of the top. Gently tap the jars against the counter a few times to remove all the air bubbles. Add more brine if needed, then seal the jars.

Once the jars cool to room temperature, open them to release any air bubbles and close them again. The pickled vegetables will last up to two months in fridge.

VINEGAR

There are quite a few ways to make vinegar. Here you will learn how to do it using alcohol or honey.

What You Need

- Alcohol that is sulfite-free and with an ABV (alcohol by volume) between 5% and 15%.
- Distilled water.
- Mother vinegar starter. You can buy it from a natural food store or use the starter from a previous batch of vinegar.
- Glass jars, sterilized with boiling water.
- Cheesecloth.
- An elastic band.
- A pot.
- A funnel.

Directions

Pour the water and alcohol into a jar using the following ratios:

- **Wine is 1:1.** You can use more or less wine to increase or decrease the sharpness of the vinegar.
- **Honey is 1:1.** Boil the water and pour it over the honey. Stir it until it is combined, then let it cool to room temperature before continuing.
- **Beer or cider.** No water.

Add the mother, then cover the jar with a cheesecloth and hold it in place with the elastic band. Store it in a in dark, warm, airy place, like your pantry. A temperature between 27C (80F) and 29C (85F) is ideal, but it will still work anywhere between 15C (59F) and 34C (94F). Leave it undisturbed for two months. It will smell.

After two months, taste test a little. Do not stick any metal in the mixture. Wood or plastic is fine. Taste it every week until you're satisfied. When it's ready, remove the gelatinous floating blob. This is the mother. Use it to start another batch.

Pasteurizing your vinegar is optional. It will make the vinegar last longer, but kill the mother, which removes some health benefits. To do it, heat the vinegar in a pot to 65C (150F). Use a thermometer. Do not let it get to 70C (160F). Once heated, allow it to cool.

Strain the vinegar through cheesecloth into a glass jar or bottle. Use a filter if you need to.

Label the jar with the:

- Type of alcohol
- Dilution ration
- Length of time fermented
- Pasteurized/unpasteurized
- Date bottled

APPLE CIDER VINEGAR

Instead of throwing out your apple cores, make apple cider vinegar out of them. You can also use all the other scraps, such as peels, seeds, stems, or any browning/discolored parts.

Using these directions, you will make apple cider vinegar (ACV) that is organic, unfiltered, and raw (unpasteurized). This means it will have all the mother which is great for gut health.

Use ACV for pickling, salad dressings, cleaning, and anything else you would normally use vinegar for.

What You Need

- Apples. Any part works. If they aren't organic apples, don't use the peels.
- 2 glass jars, sterilized with boiling water.
- An elastic band.
- Cheesecloth.
- Granulated sugar (raw, cane, muscavado, etc.) Honey will work, but not as well.
- Filtered water.

Directions

Cut the apple parts in chunks and fill the jar with them until it is three-quarters full. Cover the apple parts with filtered water, leaving 5cm (2in) of the top. Stir in the sugar until it is dissolved, then cover the jar with cheesecloth. Use the elastic band to hold it in place.

If you have some ready-made ACV with the mother (either store-bought or homemade), you can use it to jumpstart the fermentation process. Add two tablespoons of it to every liter of a new batch. Another way to speed up the process is to buy some starter yeast from a wine-making shop.

The apples must be covered with water or they will grow mold, which is any non-white scum. If you need to, weigh them down with a fermentation weight or a smaller glass jar. Make sure the weight is clean. Wrap the cheesecloth over the top of the weight, even if it sticks out of the jar.

If it grows mold at any time during the process, throw it out. The white scum is the mother and it is important for the fermentation process. Do not throw it out.

Store the jar in a dark and dry place with a temperature between 15C (60F) and 25C (80F). Stir it daily with something that is not metal, like a plastic chopstick or wooden skewer. It will bubble. That means it is fermenting, and is a good thing.

After a couple of weeks, the apples will sink. This is hard apple cider and is a little alcoholic. Strain the liquid into the other glass jar and compost the apple chunks. Do not eat them. Cover the jar with the cheesecloth and elastic band and put it in the dark, dry place. Continue to stir it every few days.

After three weeks, taste it. If it is too weak, let it ferment more. If it is too strong, add some filtered water. When you are happy with it, put the lid on the jar and place it in the fridge to stop it fermenting.

When kept in the fridge it will never go bad, but it will lose quality after a few years. If you keep it in a cool, dark place it will also last a long time, but it will continue to ferment unless you take the mother out. Taking the mother out eliminates some health qualities of the ACV.

Related Chapters:

- Fermenting

PICKLED EGGS

When you have eggs you want to preserve but powdering/drying them doesn't appeal to you, try pickling.

What You Need

- 1L jar, sterilized with boiling water
- 12 undamaged medium eggs
- 4 cups of vinegar
- 1 cup of water
- 1 teaspoon of rock salt
- Herbs and spices to taste
- A pot
- A strainer (optional)

Directions

Hard-boil the eggs then submerge them in cold water.

Boil together the vinegar, salt, sugar, and any herbs and spices you want for 10 minutes. This is your brine. Allow the brine to cool to room temperature.

Peel the eggs and put them in the jar. Pour the brine over the eggs until they are completely covered. Straining the brine is optional.

Seal the jar and shake it gently so the brine distributes evenly. Store the jar in a cool, dark, dry place. The eggs will be ready to eat in two weeks.

If the jar is unopened, the eggs will stay edible for up to six months. Once it is opened for the first time, they will last up to two weeks as long as you reseal the jar and keep it in fridge.

FERMENTING

Fermenting gives food a sour flavor without any added acid and creates good bacteria, which is fantastic for your gut health. In this section, you'll learn to ferment vegetables so you can eat them, as opposed to just making the liquid vinegar. You can ferment almost any vegetable.

What You Need

- Fresh vegetables
- A chopping board
- A knife
- Filtered water
- High-quality salt
- A glass jar, sterilized with boiling water.

Directions

Clean the vegetables and remove anything you wouldn't normally eat, like bad parts and stems. Cut the vegetables to whatever size you want and pack them in the jar. Leave at least 2.5cm (1in) room at the top. Mix one cup of water with half a tablespoon of salt until the salt dissolves. This is the brine. Pour the brine over the vegetables so they are completely covered. Weigh them down if you have to. Use a rolled leaf of cabbage or a weight on top, for example.

Seal the jar and leave it in a cool, dry area—not your fridge. After one week, taste it. If it is to your satisfaction, put it in the fridge. If you want a deeper flavor, leave it for longer. It is normal to see bubbles, white scum, or foam on top during the fermentation, but mold is bad. If you see mold, it is because the vegetable is not submerged under the brine. Discard the moldy parts. Everything under the brine should be fine.

SAUERKRAUT

Sauerkraut is fermented cabbage and a common condiment in many parts of the world. Unlike in the general fermenting process previously described, you do not need to add any water when making sauerkraut. This is because the salt draws the moisture from the cabbage.

What You Need

- Cabbage
- Carrot
- Garlic
- High-quality salt
- A knife
- A chopping board
- A bowl
- A glass jar, sterilized with boiling water
- A plastic spoon

Directions

Clean the vegetables and remove any parts you wouldn't normally eat. Shred the vegetables and put them in the bowl.

Sprinkle salt over the vegetables and knead them. This will release moisture, creating brine. Kneading is optional, but quicker. The alternative is to sprinkle the salt over the vegetables, then cover the bowl and let it sit for a few hours. When using the no-kneading method, mix the vegetables around every now and again.

Pack the vegetables into the jar. Leave a couple of centimeters of space at the top (1in) because it will continue to release moisture. Seal the jar and leave it somewhere cool and dry, but not in your fridge. Open and reseal the lid twice a day to release the gas buildup. At the same time, push the vegetables down into the brine.

If the brine is not enough to cover the cabbage after 24 hours, stir one teaspoon of salt in one cup of water and add it to the jar until the vegetables are submerged.

Taste it after two weeks. If it is to your liking, put it in the fridge. If not, leave it out for longer.

KIM CHEE

Kimchi is like the spicy Korean version of sauerkraut. This recipe will make roughly two 500g jars of kimchi, depending on the exact size of your vegetables.

What You Need

- 1 large Napa cabbage
- 1 medium radish
- 1 medium carrot
- 1 bulb of garlic
- 2.5cm (1in) of ginger
- 0.5 cup chili powder
- 1.5 tablespoons of high-quality salt
- 0.5 cup chopped scallions (spring onion)
- 2 glass jars, sterilized with boiling water
- A chopping board
- A knife
- A bowl
- A blender

All the ingredients are optional except the cabbage, salt, and at least some garlic and chili.

Directions

Wash and chop the cabbage, radish, carrots, and scallions. Discard any parts you wouldn't usually eat. Put all the good parts in the bowl.

Add the salt to the vegetables and mix it through. Cover the bowl and let it sit for a few hours. Mix the mixture around occasionally. If you want to speed the process up, knead the vegetables with your hands for a few minutes.

Blend the garlic, ginger, and chili flakes into a paste, then mix the paste in with the vegetables. Pack the mixture into the jars and push it down so the water level rises. Keep doing this until it is packed to a few cm (1in) from the top. Screw the lids on the jars and let them ferment somewhere cool and dry, but not in your fridge.

Open the jars two times a day to release the gas and push the vegetables under the liquid. Taste it in two days. If you like it, put the jars in the fridge. If not, ferment them for longer.

The kimchi will continue to ferment in the fridge, but much slower. It will stay good for at least six months in the fridge.

CANNING

Canning is a process in which food is put into a can or jar and heated to kill harmful bacteria. This heating process also pressurizes the can, and as it cools, a seal is created which prevents future contamination.

Although commonly called "canning," home canning usually involves preserving food in jars.

Depending on the exact process used, canning can preserve food for many years, but if you don't do it right, it can be dangerous. Read the following overview to help you decide which (if any) type of canning you want to try, then get more detailed instructions on how to do it depending on what you want to can.

Types of Canning

Dry canning is not technically canning. It is adding dry ingredients together in a jar as "easy-cook" meals like soups, stews, or cookies. Layer the ingredients in a jar, place an oxygen absorber on top, and screw on the lid. Label it. There are many free recipes on the internet and you can customize them as you like. Dry-canned food makes an easy homemade gift. Add a bow ribbon and a small instruction card.

Water canning is for preserving more acidic foods (those with a pH of less than 4.6). Use it for fruits, jams or jellies, pickles, and tomatoes. Making jam is a good introduction into the world of canning, as it is relatively safe. Water canning also doesn't require any special equipment besides the jars.

Pressure canning is for less acidic foods (those with a pH over 4.6). Use it for vegetables, meat, and complete cooked meals like soup, stews, and sauces. It requires a higher temperature than water canning, so you need to use a pressure canner.

Canning Equipment

You don't need much for dry canning. Any sealed jars will do. For water canning, you will need:

- A large pot.
- A rack that fits inside the pot. You can use any large pot for water canning as long as you have a rack or some other way to keep the jars off the bottom of the pot.
- A stovetop that is not glass, or an electric water-bath canner. If you do not have a the right kind of stovetop, but do not want to buy an electric water-bath canner, a propane camp stove will do the trick.
- Canning jars with lids. You must use jars specifically intended for canning. They don't have to be brand new, but they must be in good condition, without any cracks, scratches, etc. The canning lids must be new unless they are specifically made to be reusable.
- Lid bands.
- Canning salt. This is a specific type of salt. Do not try to use table salt.
- The fruits and/or vegetables you want to can.
- A jar lifter to lift the hot jars.
- A wooden spoon. Metal reacts with the canning process.
- A food strainer.
- A funnel (optional).

For pressure canning, you need a pressure canner instead of a pot and stove. All the other equipment is the same.

When canning, only can one type of thing at a time. Each item has slightly different instructions, so you don't want to get confused.

Make sure you sterilize everything before you start canning. Using boiling hot water is an easy and effective way to do it.

USDA Canning Guidelines and Recipes

Safe canning requires specific recipes. Search the internet for "USDA Canning guidelines." Do not stray from the recipes or use ones from non-reputable sources.

You also need to adjust processing time and/or pounds of pressure if you are more than 300m (1000ft) above sea level. Check the official recommendations depending on what you are canning.

Signs of Sealing

For your canned food to be safe to eat, the jars must seal properly. You may hear a popping sound as a jar seals while cooling. Once the jar is cooled, the lid should have no give when it's pressed in the center. Check this 12 hours from the start of the cooling process.

If after 12 hours the jar is not sealed, you can try to reprocess it. If it doesn't seal after a second processing, put it in the fridge and eat the contents within a few days.

Labeling

Labeling your cans is an important step. Note down what each one contains, the date and time it was processed, and the source of the recipe.

Storage

Store your home-canned food like you would any store-bought canned food. Keep it somewhere dark and dry. Do not freeze your canned food, or it will spoil.

Spoiled Canned Foods

The biggest danger with canning is botulism. Cooking the food will not make it safe. If you suspect your canned food is spoiled, throw it away without testing it. Some signs of spoiled canned foods are:

- A pop-top that does not pop when opened.
- A broken seal (the lid curves up instead of down).
- Bulging or dents near the ends.
- The food not tasting as it should.
- Leaks, cracks, or rust.
- Lids that pop in storage.
- Mold. Don't smell it, just throw it away.
- A spoiled smell. It's hard to explain the smell, but you'll know it.

A cloudy jar doesn't necessarily mean the food inside is spoiled. Boil the food for 10 minutes. If it smells bad before, during, or after cooking, throw it away.

FREEZE DRYING

Freeze-drying food is like dehydrating it by freezing as opposed to using heat. The texture will change, but the nutritional value and flavor will stay the same. It is good for long-term storage and/or camping, but doesn't work well with foods with a low water content, like leafy greens.

Fruits and vegetables are the easiest to freeze-dry. Make sure they are ripe and fresh. Wash, dry, and dice them. You can also freeze-dry meat, grains, and pasta, including full meals. Cook them as normal then allow them to cool. You must freeze-dry them as soon as they are cooled. Freeze-drying things like bread, cakes, and cookies is also possible, but it doesn't work well.

There are several ways to freeze-dry your food at home. Special machines are expensive (thousands of dollars), but guarantee the safety of the food. Use the following **DIY** methods at your own risk. If you are worried, stick to freeze-drying fruits.

Freezer Method

This is air drying inside a freezer. You will need:

- Food to freeze-dry.
- A freezer. A deep freeze works best.
- A drying rack for increased air circulation. Any tray will work if a drying rack is not available.
- Plastic wrap.
- A vacuum sealer. If you don't have a vacuum sealer, use a Ziploc bag and squeeze as much air out as you can.

Spread the food out on the drying rack or tray and cover it with plastic wrap. Put it in the freezer. After three weeks, bring once piece to room temperature. If it changes color, it's not ready. Leave it another week before testing it again.

Once the food is ready, vacuum-seal it or store it in a Ziploc bag. Store the freeze-dried food in cool, dry, dark place.

Dry Ice Method

Using dry ice is faster than the freezer method. Do it on a low-humidity day and wear the appropriate clothing to prevent getting burnt by the dry ice. You will need:

- Food to freeze-dry.
- Insulated gloves.
- Long-sleeved clothing.
- An insulated container twice the size of the amount of food you want to dry.
- A vacuum sealer or Ziploc bags.
- Dry ice at a 1-to-1 weight ratio with the food.

Vacuum-seal the food or place it in a Ziploc bag. Squeeze as much of the air out as you can. Place the bagged food in a single layer in the bottom of the insulated container and cover it with the dry ice. You can stack layers of food and ice. Leave the lid open a little to allow gas to escape. Once the dry ice is gone, the drying is done. It will take approximately 24 hours. Store the freeze-dried food in cool, dry, dark place.

Reconstitution

Berries and apples don't need to be reconstituted. Just eat them as they are. For everything else, put the food in a bowl and add boiling water bit by bit. It will begin to reconstitute. Add more water as needed.

Related Chapters:

- Drying Fruit

ROOT CELLAR

A root cellar is a storage unit built underground where the temperature is cooler and more consistent. It is a great way to store what you grow in the summer so you can enjoy it throughout winter. It is not suitable for meat or dairy products.

In a year-round warm climate, like in the tropics, a root cellar will not work well, but you can grow fresh produce all year round.

You can construct a small root cellar using any suitable container such as a large trash can or Esky cooler. Make sure it's made of non-corrodible material such as galvanized steel or heavy-duty plastic.

If you want to store a lot or several different types of food but don't want a big construction project, make several of these small ones and store things separately for convenience.

Not all types of fruits and vegetables will store well in your root cellar. Some common ones that do are apples, beets, broccoli, Brussel sprouts, cabbage, carrots, leeks, parsnips, pears, potatoes, turnips, radish, beans, garlic, onions, pumpkins, squash, sweet potatoes, and tomatoes. The list isn't comprehensive. Research what you want to see if it will store well or not.

Another thing to consider is that not all foods store well together. Keep fruits and vegetables separate. Sweet potato and potato must also be kept separate.

You can store these together but separate from everything else:

- Peas and onions.
- Pumpkin and squash
- Apples and pears

In general, everything else can be stored together, but do your research first to make sure.

What You Need

- Containers
- A drill
- A shovel
- Rocks
- Straw
- Fresh produce, undamaged and unwashed. You can brush it, but don't use water.
- Plywood

Directions

Choose where to put your root cellar. You want the ground to be:

- Easy to access and not too far from your home.
- Not covering existing drainage lines.
- Sandy, if possible, as that will make it easy to dig in and better for drainage.
- Shaded to promote a cool temperature.
- Slightly elevated for drainage.

For larger projects, you also need the ability to dig a drainage line, such as a French drain, around the cellar. This is not an issue with this small version.

Once you know where you want to put the root cellar, dig a hole to put your container in. It must be about 0.5m (1.5ft) deeper than the height of your container. You need enough space to add a layer of rocks for drainage and to ensure that once it is inside the hole, the top of the container is below the frost line. Once the hole is dug, throw a layer of rocks in the bottom.

Drill a few holes in the bottom of the container for drainage and put it in the hole. Layer straw and food inside it, starting with straw. The food must be in single layers. Once it the container is full, put the lid on, followed by a layer of straw, a plywood cover, and finally a rock to weight it down.

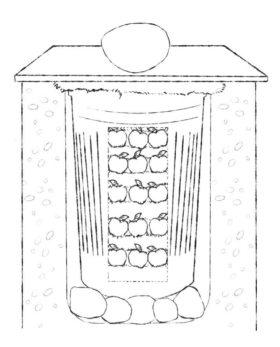

If your fruits or vegetables start to shrivel, you need more humidity. Spray mist onto the straw.

If the food has condensation or moisture, there is too much humidity. Make sure there is no water underneath your container and replace the straw.

COOKING

In this section, you will learn several ways to cook without your stove. Combine all of them, and you will never have to cook with gas or electricity again. For example, cook things in the solar oven during the day and then put them in the thermal cooker for later. Have a mud oven and/or rocket stove as a backup.

Depending on what country you live in, many homes already have wood stoves built into the house. If not, you may want to get one professionally installed. They are a great way to heat a home and you can cook on the. Make sure you use cast iron cookware, such as a Dutch oven.

SOLAR OVEN

A solar oven uses nothing but the sun's energy to cook food. It is a slow cooker that is good for cooking soups, stews, rice, and bread. You can also use it as a food dehydrator.

Here is how you can make one out of a cardboard box and aluminum foil. If you like it, you can make or buy a studier and/or larger one. On a sunny day, this solar oven can get up to 150C (300F), so use oven mitts and sunglasses, and keep your pets and children away from it.

What You Need

- A pot with a lid. It should be dark-colored, large enough to cook what you want, and lightweight. If you don't have a lid, paint some aluminum foil black and cover the pot with it.
- A Styrofoam cooler big enough to hold your pot.
- A marker.
- A ruler.
- A knife.
- Tape.
- Aluminum foil.
- Black paint.
- Heavy-duty clear plastic or plexiglass.
- Cardboard.
- Duct tape.
- A small cooking rack, such as a cake rack.
- A drawing pin.

Directions

Place your pot in the cooler and cut the Styrofoam so it angles down while leaving enough room for your pot. Cut it from the inside edge.

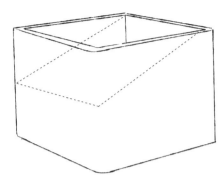

Cover the cut edges with tape to prevent the Styrofoam from fraying.

Line the inside with aluminum foil, shiny side out. Tape the aluminum foil down and fold it to size. Be careful you don't tear it. Make sure you use tape and not glue so it is easier to replace it when the time comes. Paint the bottom black.

Cut the heavy-duty clear plastic the same width as the opening of the cooler. Length-wise, you need to leave some going over the edge so you can pin it in place. If you're using plexiglass, it needs to be the exact size of the opening. The weight will hold it in place.

Make four reflectors from cardboard and line them with aluminum foil. See the image for the wing shape. The inside flap needs to be the exact width of the edge of the inside of the cooler.

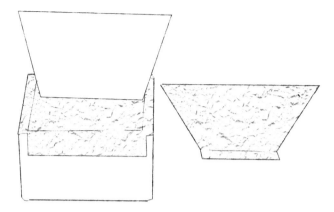

Measure each wing separately since the diagonal lengths will be different. Making them wider angles means they will capture the sunlight for longer. Tape three of the wing corners together with duct tape.

Once the fourth panel is in place, you can use more tape to reinforce it. Leave one corner un-taped. You can hold it in place with a couple of holes and a twist tie, but that is optional.

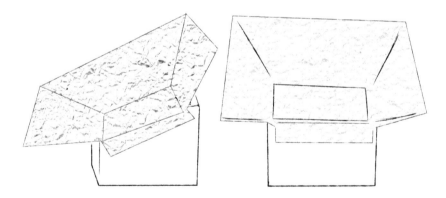

Put the rack inside.

Variations

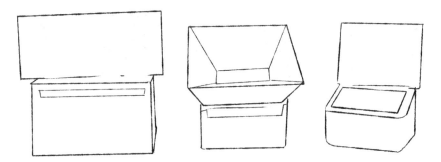

Cooking

Preheat your solar oven by setting it out in the sun 30 minutes before putting the food in it. Put your food in it and let it cook.

The smaller the shadow, the more sun the oven is getting. Turn it throughout the day to keep the shadow as small as you can. The wider the reflectors, the less you will need to move it.

Here are some estimated cooking times. Expect things to take longer on cloudy days.

- 2 hours: Eggs, fish, chicken (sliced), above ground vegetables, fruits.
- 4 hours: Root vegetables, lentils, small pieces of meat.
- 8 hours: Roasts, soups, stews, large dried beans.

THERMAL COOKER

A thermal cooker (also called a wonderbox or haybox) is a way to continue cooking food once it has been heated up. For example, you can use it if the weather turns bad while you're using your solar oven or to conserve fuel.

It is an insulated box that works similarly to a thermos. It will keep things hot or cold for an extended time. Once the food is heated, you put it in your thermal cooker and let it sit there. It takes a while to cook, but you do not need to monitor it. The food will never burn, and it is not a fire hazard.

What You Need

- A pot. Using the same one as in your solar oven is a good idea.
- A Styrofoam cooler big enough to hold your pot with about 10cm (4in) clearance on each side.
- Insulation materials. You can use blankets, dry hay, wood chips, Styrofoam, sawdust, scrunched-up newspaper, etc.

Directions

Line the bottom of the cooler with insulation 10cm (4in) thick. Put your heated pot inside and surround it with more insulation. Pack it as tightly as you can. Have at least 10cm (4in) of insulation on top.

For best results, boil the food for at least five minutes before putting it in your thermal cooker. Large beans need to boil for at least 15 minutes. It will take a minimum of several hours to finish cooking the food.

If you are good with a sewing machine, you can make a neat and pretty one out of heavy fabric and bean bag beans. Search the internet for "wonderbox pattern."

PORTABLE ROCKET STOVE

A rocket stove is an efficient wood-burning stove. This portable version is powered with twigs, and although it is primarily for cooking, it is an efficient source of heat too.

What You Need

- 1 large can with a lid, such as a large paint can.
- 1 medium can, such as a smaller paint can.
- 1 small can, such as a food can.
- Gloves.
- Paper.
- Pencil.
- Scissors.
- Tin shears.
- A can opener.
- Pliers.
- Fire-resistant insulation, such as dry wood ash, perlite (hydroponic soil), vermiculite, or pumice. You can use sand if you have nothing else.
- A small cake rack.

Cut metal is sharp. Make sure you wear the gloves.

Directions

Put the small can on a piece of paper and trace the bottom of it as a template. Cut it out and use it to cut the same size holes near the bottoms of the medium and large cans. Make the hole in the medium can a touch lower.

Cut off the bottom of the small can with the can opener, then cut four evenly spaced slits in the bottom of the can a couple of cm (1in) deep.

Line the bottom of the large can with some fire-resistant insulation material.

Put the medium can inside the large one and line the holes up.

Slide the small can, slits facing in, through both holes. Bend the flaps made by the splits to keep the small can in place. Fill all the space between the large and medium cans with the insulation.

Cut a hole out of the large can lid big enough to expose the top of the medium can. Place it on top of the large can.

With the part of the lid you cut out, make a little shelf to put inside the small can. This is for airflow.

Place the small cake rack on top of the large can. If you don't have a suitable cake rack, use anything heat-resistant that will create a small gap between the pot and the top of the medium can.

Build your fire on the top shelf inside the small can like you would any other fire. That is, start with the small stuff first and gradually building to larger twigs.

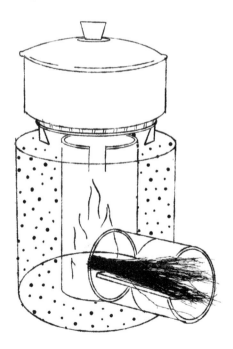

MUD OVEN

Mud ovens are used all over the world in some way. Some of the many names they go by are adobe oven, clay oven, earth oven, el horno and more. All these types may use different materials and/or shapes, but they work in similar ways.

A fire is lit inside the oven and burnt down to coals. Then the coals are removed and the food is placed inside to cook slowly. There is almost no chance of burning your food when cooking with a mud oven.

Mud ovens are fuel efficient and retain heat for up to 24 hours when their doors are kept closed. They stay hot on the inside but won't burn if you touch the outside.

This project may seem a bit daunting if you are not naturally handy, but all you will lose is your time. If you mess it up, just try again with the same materials.

Here are some basic instructions to give you an idea of what's involved when building a mud oven. If you want to make one, you may want to get more detailed instructions off the internet.

What You Need

- A shovel
- Water
- Chicken wire (optional)
- Sand
- Straw (optional)
- A tarpaulin
- Bricks or stones
- Empty glass bottles
- Fire-resistant insulation, such as perlite (hydroponic soil)
- Newspaper
- A metal spoon

- Paper
- Hardwood
- Aluminum foil (optional)
- Matches or a lighter

Planning

Plan your mud oven so it is just big enough to hold what you want, such as a large pizza, a couple of loaves of bread, or a couple of cooking pots. Keeping it small helps with fuel efficiency and consistency in heat distribution.

After building the basic dome oven, you may want to shape it into a sculpture or something practical, like a bench. This will require additional materials, but if you want to do it, then make sure you put it somewhere with enough space.

You may also want room to build a roof to protect the oven from rain, but this is optional.

Here are some basic calculations to give you an idea of size:

- Door width = Can fit your largest object (a 15-inch pizza, for example) in
- Base width = Door width x 3
- Oven width = Door width x 2.5
- Oven height = (Base width / 2) + 10cm (4in)
- Door height = 0.63 x oven height (63%)

Once you know how much room you need, you can decide where to put the oven. Choose wisely, because it is a permanent structure. Consider the distance from your home for safety and convenience, the wind direction, and the gradient of the ground. It will be easiest to build it on level ground, but that is not 100% necessary.

Find Clay

Now you know where to build your oven, find the closest source of clay you can. It does not have to be pure clay, but it needs to have a sticky texture so you can build with it.

Dig in the ground below the topsoil near water sources and where there are cracks in the earth. Clay can also be in random places, so poke around a little.

When you think you found some, test it by mixing a handful with a little water and kneading it to the consistency of modeling clay. Put it through the following tests:

1. If it sticks to your hands, it probably has clay. Move on to test two.
2. Flatten it to the thickness of a pencil. If it wraps around your finger without too much cracking, move onto test three.
3. Roll it into a small ball and allow it to dry. If you can squeeze it when it's dry without it crumbling, it is good to use.

Once you have found clay, you need to mine it. If it is too hard to dig, soak it with water first. Sift it through chicken wire if there are a lot of rocks and/or sticks in it.

Only dig out the amount of clay you need for the stage of the project you are on, and consider the environmental impact of mining it. It's possible the hole will turn into a pond, for example. Perhaps you can make it into something useful or fill it with land from somewhere else.

Prepare the Clay for Building

Mix the clay with water until it is at a consistency you are happy to work with. Every clay composition is different so there are no specifics, but you want it pliable enough to mold while strong enough to hold its shape.

Consider:

- The wetter it is, the longer it will take to dry.
- If it's too wet, it won't hold its shape.
- If it's too dry, it will be too hard to mold.
- The weather you are working in will affect how long the clay takes to dry.

For a standard soft clay mix, put the clay in a bucket and soak it in water for a few hours.

If you want something softer, stir it up a bit. If it gets too sloppy, add more clay. To make something stiffer, add sand and straw and mix it in on a tarp by stomping on it. This is good for additional designs, especially if they are for things that will bear weight, such as a sculpture or bench.

Create the Base

This is the stand that you will build your mud oven on. You can build it out of anything fireproof (rocks or bricks, for example). Build it waist height so you don't have to bend over to access your oven.

Dig a circular hole 20cm (8in) deep. The diameter of the hole should match the width of the base of your oven.

Build a ring of stones/bricks to waist height, preferably without using mortar. Fill the ring with empty glass bottles laid on their sides and point their necks toward the center. This will create insulation. Fill in the gaps between the bottles with any other type of fire-resistant insulation.

Finish off the base with a 15cm (6in) layer of clay on top. Make it level and allow it to dry completely before moving on to the next step. This last layer of clay is the oven floor.

Make a Sand Mold

Mix sand with a small amount of clay and water, and then use it to build a dome in the center of the oven floor. Use the oven and width height calculations from the planning stage. Make sure the edges of the sand are at least 15cm (6in) away from the edges of the oven floor.

Once the sand is mostly dry, move on to the next step.

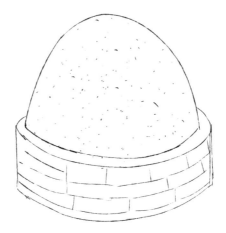

Build the Oven

Cover the sand mold with wet newspaper. Make sure the newspaper is flat with no creases.

Starting from the base and working your way up, cover the newspaper with clay 10cm (4in) to 15cm (5in) thick. Ensure each clump you put on blends with the others so there are no seams. Make the clay even and smooth it over with your hands.

Don't worry about leaving space for the door. You will cut that out later.

Let the clay dry until you can poke it with your finger without leaving a dent. This can take days or weeks, depending on the weather. Erect a tarpaulin roof above it to protect it from the rain, or build a permanent one if that is your plan.

Remove the Sand

Once the mud oven is dry, mark out the door by scratching it into the clay.

Cut a hole in the door area just big enough to put your hand in and remove a narrow channel of the sand.

Carve out the rest of the door and dig out the rest of the sand. Stop when you hit the newspaper. Make the entry nice and smooth.

Finish the Outside

Rub the exterior of the oven with the back of a metal spoon to finish it. You can add sloppy clay to fill in holes where it is needed, but do not paint or cover it with anything that will keep it from breathing.

Make the Door

Make a template of the door opening from paper, then cut that shape out of a piece of hardwood, such as redwood. It doesn't need to be a perfect fit, but the closer you get it, the better.

Line the interior of the door with aluminum foil or soak it in water overnight to keep it from burning. Make a wooden handle as well.

Proof Your Oven

Build a fire in your oven and keep it burning for eight hours. This will harden it.

Usage

To cook in your mud oven, build a fire in it and leave the door off. After a few hours, the black soot on the inside will disappear. This means it is ready to use.

Remove the coals and put whatever you want to bake inside. Put the door on to retain the heat.

WATER

The need to procure your own water depends largely on where you live, but even if you have a reliable government-sustained water supply, it is a good idea to have at least one alternative water source, in case a water pipe bursts, for example.

Another reason to do it is so you have more control over what is in your water. Collecting and treating your own often produces cleaner water than the pipe-and-chemical system governments use.

If you live off-grid, have at least two ways of procuring water. The more the better.

All water collection methods involve the following stages:

- Collection
- Filtration
- Channeling
- Storage
- Treatment
- Distribution

The stages may not happen in this order and can be inherent in the system.

Collection

Collection is when you get the water from a source, such as rain, a well, fog, springs, lakes, rivers, or streams.

This section has instructions on using the first three sources. If you want to use the other ones, you need to create a way to get the water from the sources to your home—by channeling, for instance.

You must also consider water rights. Tapping natural water sources that do not originate on your property may result in legal problems. Check what the diversion rights are and if you are allowed to take the water.

Filtration

Filtration removes larger contaminants such as leaves, sticks, and rocks. A grate or mesh screen is usually sufficient for this.

Channeling

Channeling is how you get the water from the collection point to your storage area. You can do it manually—with a bucket, for example—or via piping or gutters.

Channeling with a gutter system usually relies on gravity.

When channeling via piping, you will use either gravity or a pump system. When using a pump, it is best to use a manual one for sustainability. Even if you use an electric one, you should have a manual backup in case the power goes out. Consider a RAM pump.

In cold climates, make sure any piping is insulated to prevent it from freezing.

Storage

This is where you put the water you collect. There are many options, like 55-gallon drums, IBC containers, or larger water tanks, and you can plumb several storage units together to scale them. When doing this, don't stack them on top of each other. Build stands instead.

If you live in a cold climate, make sure you drill an overflow hole at the 90% full mark. This is because frozen water expands, and if your tank is full when the water freezes, it will get damaged.

Only use tanks approved for water storage that have had no chemicals in them.

Treatment

Water treatment makes the water useable. The amount of treatment needed depends on how dirty the water source is and what you want to use the water for. If the water is just for your garden, then it won't need much (if any) treatment, but you still want it clean to grow things you eat. Water for hygiene and consumption (potable water) must be treated to standards. Keep potable and non-potable water separate and well labelled.

The most basic, reliable way to treat water is to boil it for five minutes, but this is not really a practical long-term solution.

It is possible to get "whole house" filtration, in which the water is filtered before it enters your house (between your well and water faucets, for example), but this is expensive to do. A more accessible option is to get gravity or in-line water filters. Berkey and Aquarain are two reputable brands, but there are others. Whatever you choose, make sure the filters are quality-tested and follow the manufacturer's instructions.

Distribution

Distribution is getting the water from storage to wherever you want to use it. Often this is just a tap and/or piping system running from your water storage tank.

If you want to hook it up so you have running water in your home, you need constant water pressure. The easiest way to do this is with a large water-pressure tank, which is also a storage tank. Pump (or gravity-feed) the water from the collection point into the tank, then from the tank to your faucets.

RAIN BARREL

For most people, collecting rainwater is the most viable option for creating a water supply.

Depending on the design of your home, your roof is likely the catchment area and your gutters are the channeling system. Direct your gutters so the water flows into a storage vessel. Alternatively, or in addition, hang a tarpaulin between some trees and funnel the rainwater straight into storage.

This simple rain-barrel design can be used with almost any home with a gutter system, and if you want to scale it up, you can just plumb a few of the buckets together.

What You Need

- A large plastic rubbish bin. Clean it well. This will be the barrel.
- A spigot.
- A drill.
- Sealants such as rubber washers or plumber's tape.
- A hose or PVC pipe.
- Landscaping fabric.

Directions

Drill a hole near the bottom of the bin for the spigot. Insert the spigot and seal it so there are no leaks.

Cut a hole in the bin lid large enough so the water can flow from your gutter down-pipe into it.

Make a third hole near the top of the bin that is the same size as the hose or PVC pipe. This is the overflow spout. Use some hose or PVC pipe to either connect it to a second barrel or divert the water into the earth. This will prevent water damage to your home.

Cover the top of the barrel with landscaping fabric and place the lid on top. The landscaping fabric is a filter. Water can get in, but nothing else.

Put the rain barrel directly under your downpipe but off the ground. Place it on top of a few cinder blocks, for example. Do not put it near any septic or utility lines.

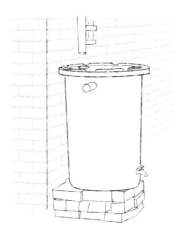

Improvements

The above is a basic design. Here are some small improvements you can make depending on your needs:

- Attach the gutter directly to the rain barrel with a downspout adapter. This will remove the need for the fabric filter and capture more water.
- Raise it high enough to make it more convenient to use the tap.
- Attach an inline filter on the inside, where the spigot is, for potable water straight from the barrel.
- Have two spigots, one potable and one not. Label them well.
- Use a bigger and/or multiple barrel(s).
- Attach the barrel directly to your garden's irrigation system.

- Divert all gutters to the rain barrel or make a rain barrel system for each downpipe.
- Have a debris filter on top of the downpipe.

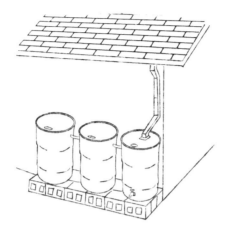

Tarpaulin Adaption

When you want to collect rainwater where there is no roof and guttering system, you can set up a tarpaulin.

In this example, the tarpaulin is strung up between trees and shaped into a funnel. The water is channeled from the tarpaulin directly into a storage tank (in this case an IBC container) via a PVC pipe.

SHALLOW WELL

Digging a shallow well is a great way to tap water for irrigation. You may be able to drink it, as long as it passes testing, though deeper wells are better for drinking water as they have more natural filtration.

To dig a shallow well, you need to have a water table 7.5m (25ft) or less below the earth's surface. You can find out your water table levels from your local council or by asking other local residents. While you are at it, make sure it is legal to dig a well in your area.

This method uses a hand auger. It assumes you have soil that is easy to dig in. If you have difficult soil you will need an alternate way to dig, such as well-point, water drills, or other mechanical means. The basic steps are still the same, so at least you will have an idea if you want to attempt it yourself, hire someone to do it, or not bother at all.

What You Need

- A hand auger and extensions.
- A shovel.
- 4 bags of pea gravel.
- 1 bag of bentonite (or quickrete or similar).
- A 1.25-inch brass foot valve.
- PVC primer.
- PVC cement.
- Teflon tape.
- A pitcher pump. Make sure it is certified lead-free.
- A wooden board.
- Screws.
- A bucket.
- A string and a nut.
- A measuring tape.
- Wrenches.

- A saw.
- An electric grinder (optional).

For the casing pipe:

- 10m of PVC pipe of a minimum 4in in diameter.
- A PVC cap to fit the pipe.
- A PVC flange to fit the pipe.

The length of the PVC pipe needs to be the depth of the well plus 1m (3.3ft). Buy 10m (33ft) and cut off the extra when you know the exact depth.

The larger the diameter of the casing pipe, the more water you'll be able to "store" for immediate use. It must be smaller than the hand auger, with enough room to add some pea gravel.

For the water pipe:

- 1in PVC.
- 1in couplers if needed.
- 2 x 1in to 1.25in PVC threaded adapters.

The length of the 1in PVC pipe is the length of the casing pipe minus 30cm (1ft).

Couplers are for joining pieces of PVC pipe together. If you have a choice, always get the longer couplings.

Joining PVC

Use this gluing method for all instances of connecting PVC so you don't get leaks.

Apply lots of primer to both sides of the PVC pieces you want to join, then apply a coat of pipe cement where you put the primer. Immediately connect the two pieces and hold them together for 15 seconds. Allow the pipe to rest a few minutes more to dry before working with it.

Where to Dig

Putting your well in the right spot is very important. You need to find where there is water with a good reproduction rate. Here are some tips:

- Heavy sand and gravel deposits are a good sign.
- Low points in the land, like valleys, often contain water.
- Spots near natural bodies of water are likely to be productive. If there's no vegetation, that means no water.
- Use groundwater and/or topographic maps.

Other things to consider when you're placing your well are:

- Its proximity to your home or wherever you want to use the water. Keep it a minimum 15m (50ft) from any contaminants, such as septic tanks, muddy areas, sewers, or animal pens.
- The need not to hit any existing utility lines.

Dig the Hole

Depending on the ground, digging with a hand auger isn't hard, but it isn't easy either. It is a good idea to get someone to help you.

The main thing is to make sure you dig straight down. Bring up the dirt each time and add extensions as you need to. Once you get near the water table, you will get suction. Use a twist-and-pull motion to get the auger unstuck.

There is usually water on the other side of clay. The color of the dirt you bring up will give you clues.

The deeper you dig into the water table, the more water you will get, but you need to weigh this against the extra labor that is needed to dig.

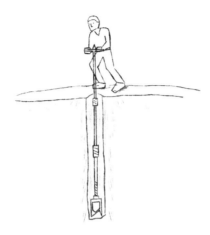

Once you finish digging, lower some string with a nut tied to it down the hole until it hits the bottom. Tie a knot in the string at ground level so you know how deep the hole is.

Pull it back up and measure the length of string that is wet. That is the depth of the water table.

Create a Well Screen

A well screen creates a larger surface area for water to seep in while keeping larger contaminants out. To make it, cut a bunch of thin slits into the bottom of the casing pipe.

You can use a hacksaw or an electric grinder with a thin disc to make the slits. Another alternative is to drill lots of holes with small drill bit, though thin slits are better.

Make the first slit 15cm (6in) from bottom of casing pipe. Create three of them, a few centimeters (1in) apart.

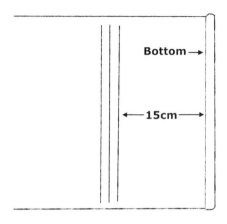

Do this every 5cm (2in) up the pipe until you get to where the top of the water table would sit.

On the other side of the pipe, repeat the process, but put the slits so they are between the other ones. There will be three 15cm slits every 2.5cm on the pipe, with groups of three on alternating sides of the pipe.

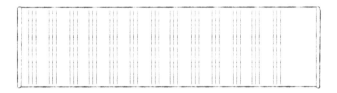

Attach the PVC cap to the end you left the 15cm (6in) gap on.

Install the Well Casing PVC

Put the PVC pipe, well screen first, into the hole. Surround it with pea gravel until the gravel is higher than the well screen slits.

Fill the rest of the hole around the pipe with dirt. Use the dirt you pulled out while digging the well and put it back how it was originally.

Install the Water Pipe PVC

Use the string and nut to measure from the top of the casing pipe to the bottom of the well. Make sure your water pipe is this length minus 30cm (1ft). You do not want the well pipe to touch the bottom of the casing pipe.

Attach the brass foot valve to the bottom of your water pipe. This lets water in, but not out. Use the PVC threaded adapter to attach it. Put Teflon tape over the threads, but do not glue it.

Make the Pump Stand

Attach the PVC flange to the wooden board. You will need to drill it. Drill a 1in hole in the center for the well pipe as well.

Sit the pump stand on top of the casing pipe.

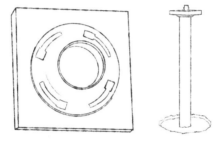

Attach the Pitcher Pump

Fill the water pipe with water from another source. This primes the well. Once it overflows, attach the pitcher pump to the pump stand.

Pump the pump until the dirty water clears. This may take a while.

Well Cap

This is a seal where the well casing meets the ground. It will prevent contamination of the water table.

Dig 30cm (1ft) around the casing and 20cm (8in) deep. Fill the hole with bentonite (or whatever you are using). Make it higher at the pipe and tapering down so it will shed water. Once it is dry, you can cover it with soil, though this is optional.

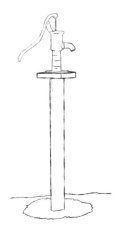

Usage

Keep some water near your new well until you are confident that it will continue to prime on its own. If it doesn't prime (or stops priming), it is probably because the foot valve is dirty. Take the valve out and clean it.

If you plan on drinking the water, make sure you test it. It may not even be suitable for plants and animals.

You can pump water from your well to a storage tank. If you want to use an automatic pump to do it, make sure the GPM (gallons per minute) of your pump does not exceed the GPM of your well. It is also a good idea to make the pump shut off automatically when the tank is near full and/or to run it at timed intervals to give the well time to replenish.

FOG/MIST HARVESTING

If you live somewhere that gets a lot of fog/mist, you can get fresh water via fog harvesting. Fog harvesting works best with persistent fog and heavy winds and is great to use in conjunction with rainwater collection, since fog often comes when there is no rain.

This method uses a mesh netting that hangs in the fog and condenses it into water droplets, which you channel into your water storage. The sun desalinates the water so it is good to drink after treatment.

What You Need

- PVC piping or any other non-corrosive construction material, such as bamboo or wood
- A water storage container, like an IBC tank or water drum
- Rachel mesh
- Twine
- A funnel
- A hose

Different mesh yields different results. Rachel mesh works well for its price. There are special meshes out there if you don't mind spending more and want to hunt them down.

Directions

Find some firm, flat ground that gets fog and is close to where you want to use the water.

Decide how high you want to hang the mesh. You want it so the mesh hangs in the middle of the fog. Your support posts need to be 125% that height if the mesh will be freestanding so you can dig them in deep enough to support themselves.

Dig the posts into the ground and space them a little closer together than the width of your mesh. Make sure you do it so the mesh will be facing the wind that brings the fog. Attach guy-lines for more stability if you need to.

Stretch the mesh out between the support posts and attach them with twine, wire, or zip ties. You can create a diagonal support across the mesh if you want, but that is optional.

Create a gravity-fed gutter system to collect the water droplets and direct them into your water storage container. Use half a PVC pipe or a rain gutter and attach it to the bottom of the net and the support posts. Make sure it angles down slightly.

Attach the lower side of the gutter to a funnel and hose system to minimize loss and contaminants. Feed the other end of the hose into your water storage container.

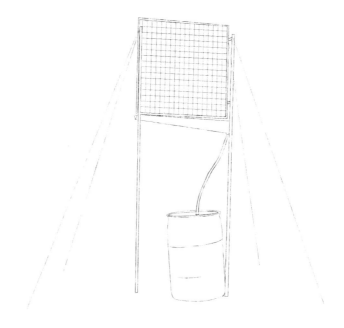

WATER TESTING

It is important to regularly test the water from any source you plan to drink, bathe, or cook with.

When to Test

- Upon installation of a new system or after maintaining an existing system.
- Annually for any source, at a minimum.
- Seasonally (four times a year) with a shallow well or surface water source.
- If your well runs dry and comes back.
- If the water changes in smell, taste, or color.
- When you are expecting a baby.

How to Test

Use a water-quality testing kit that can test for bacteria, lead, pesticides, nitrites, nitrates, chlorine, hardness, and pH. Follow the manufacturer's directions. The kit should come with safe limits for certain contaminates. If it doesn't, use these as a guideline. These are maximum safe limits.

- Bacteria - 0
- Nitrate nitrogen - 10 mg/L
- Nitrite nitrogen - 1 mg/L
- Arsenic - 10 ug/L
- Radon - 4,000 pCi/L
- Uranium - 20ug/L
- Lead - 10 ug/L

If you want to drink and/or bathe in the water, consider getting it lab-tested. Use DIY testing first to see if you should bother.

Related Chapters:

- Shallow Well

GRAVITY CERAMIC FILTER

A ceramic filter is a reliable way to treat water so it is safe to drink. The porous ceramic has many tiny holes that filter out contaminants as water passes through it.

You can't make the ceramic filter element, but sourcing one is not an issue in most places, and you can store unused filters for a long time.

In this project you will use a ceramic filter element to create a gravity filter. This is cheaper than buying a commercial gravity filter, and you can adapt it to what you need you need. You can put one in your rain barrel, for example.

What You Need

- A candle or pot ceramic filter element. Ensure it meets drinkable standards.
- 2 food-grade buckets.
- A spigot.
- A drill.
- Cheesecloth.

Directions

Bucket 1 is the top bucket. This is where you will put the unfiltered water. Bucket 2 stores the filtered water.

Create a hole in the center of the bottom of bucket 1 to fit the ceramic filter, then attach the filter so the nipple goes through the hole. It must fit flush, so no water can sneak through without passing through the filter.

Cover the filter with the cheesecloth to trap larger particles. This will increase the life of the filter. It is not shown in the image.

Put a hole in the lid of bucket 2 so it lines up exactly with the hole in the bottom of bucket 1. Drill a hole 5cm (2in) from the bottom of bucket 2 to fit the spigot. Follow the manufacturer's instructions to fit the spigot.

Place bucket 1 on top of bucket 2. The filter's nipple needs to pass through bucket 2's lid. Pour water into bucket 1. It will pass through the filter into bucket 2, and you can then drink it. It will take a while to filter.

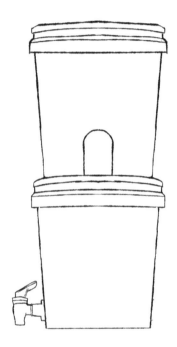

Maintain the filter as per the manufacturer's instructions.

Related Chapters:

- Rain Barrel

SODIS

The SODIS method uses the UV rays from sunlight to treat water. This is an effective, chemical free, and low-cost way to create drinking water.

The problem with the SODIS method is that it is weather-dependent and slow. On a sunny day, which is ideal, it takes six hours to treat a 2L bottle of water.

SODIS works best in places within 35 degrees of the equator.

What You Need

- A transparent PET bottle no larger than 2L (0.5gal).
- Clear water.

The bottle must be clean and damage-free. Undamaged and uncolored soda bottles are popular choices.

Use the clearest water you can find. It must be clear enough to count your fingers on the other side once the water is in the bottle. If you don't have water that clear, you can run it through a fine cloth and/or a bio-filter.

Directions

Wash the bottle for the first use and fill it three-quarters full with water. Put the lid on, shake it vigorously for 20 seconds, then fill it the rest of the way to the top.

Check it for clarity and leave it in the sunlight for the required amount of time:

- Sunny = 6 hours
- Partly Cloudy = 24 hours
- Very cloudy = 48 hours

- Raining = Not effective. Drink the rainwater instead.

You can maximize the sunlight by putting the bottle(s) on a reflective surface, like aluminum foil or a metal sheet, and/or sloping it/them towards the sun.

Once the water is ready, store or drink it straight from the bottle. Transferring it increases the chance of contamination.

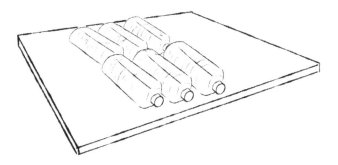

BIO SAND FILTER

A bio sand filter uses sand, gravel, and a natural biological process to filter out contaminants. Once built, it takes a minimum of 10 days to produce drinking water.

The components of a bio sand filter are as follows, from the top down:

- **Lid**. Prevents contaminants from entering the water.
- **Reservoir**. Holds unfiltered water.
- **Diffuser**. Spreads the water evenly, minimizes disturbance to the bio-layer, and keeps large particles from entering the system.
- **Resting water level**. Prevents the sand from drying out, which helps create the bio-layer.
- **Bio-layer**. Good bacteria to clean the water. It develops in the top 5cm (2in) to 10cm (4in) of sand.
- **Filtration sand**. Removes more contaminants. It needs to be 0.15mm in particle size and 15cm in depth.
- **Separation gravel**. Prevents sand from going into the outlet tube.
- **Drainage gravel**. Supports the first layer of gravel and helps keep sand from going into outlet tube.
- **Outlet**. Transports the clean water to the clean water container.
- **Clean water container**. Stores the clean water ready for use.

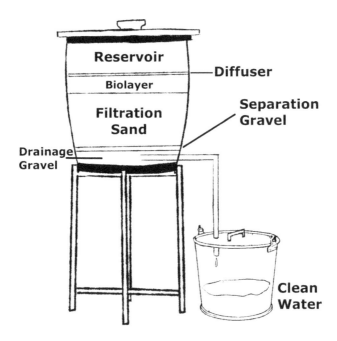

These instructions are an adaption of the HydrAid Bio Sand Filter:

https://www.youtube.com/watch?v=grQX4v4E9Ao

What You Need

- Bricks.
- A 20L (5gal) food-grade plastic drum with lid.
- 1 extra lid or other food-grade piece of flat plastic for the diffuser plate.
- PVC piping, joiners, and end caps.
- A drill.
- A 2mm drill bit.
- Tubing.
- Rubber grommets.
- A food-grade bucket with a lid.
- River rocks.
- Pea gravel.

- Fine sand.
- Water. Use the cleanest non-potable water you have available.

Wash everything well, including the gravel and sand.

Directions

Find somewhere flat and out of the way that is still convenient to access. Build a low stand from bricks and place the drum on it. This is its permanent position. Moving it will disturb the filtration layers.

To make the outlet, drill a bunch of small holes in a couple of short pieces of PVC pipe and join them together in parallel. Make sure they fit inside the drum. The holes let the water exit the filter without clogging the outlet tube. Clean any plastic bits out of the pipes and attach tubing to transport the water out of the filter.

Drill a hole in the bottom of the drum for the tubing to pass through and use the grommets so it doesn't leak. Plug the tubing during the setup phase. Test it to make sure there are no leaks.

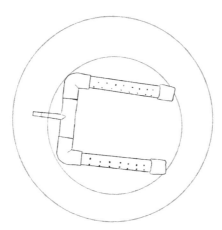

Use the flat piece of food-grade plastic to make the diffuser plate. Cut it so it fits snug and flat inside the drum at the right place. There must be 5cm (2in) of space between the sand and the diffuser for the bio-layer to form. There must also be enough room between the diffuser and top of the drum to add unfiltered water.

Once you have the size right, drill a bunch of small holes in the plate.

Mock fit the diffuser where you want it to go, and test it. Mark a fill line for the sand in the drum.

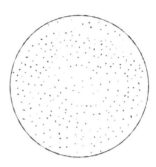

Fill a quarter of the drum with water, then cover the outlet with an even layer of river rocks. Add a 3cm- (1 in) layer of pea gravel on top of the river rocks, then add more water.

Add the filtration sand up to the line you made before. Add more water as needed so the sand is always entering water, and pour the water through the diffuser to help keep everything even. As you add the sand, use your hand to help spread it out.

Set the diffuser plate in place, then fill the drum up to the top with water and put the lid on.

Unplug the tubing so water will come out of it. Direct it into an empty bucket, but not into your clean water container.

For the first couple of days you need to add water continuously, until all the components settle. If sand settles more than 5cm (2in) below your sand mark, add more sand. Always keep at least 5cm (2in) of water above the sand.

It takes a minimum of 10 days for the bio-layer to form. Do not drink the water before then, but feel free to use it on your garden.

Do not add any chemicals (purification or otherwise) to the filter. If you want to treat it further, add the treatment to the post-filtered water.

After 10 days, test the water to make sure is it good to drink. If it is, direct the tube outlet to your clean water bucket. Create a hole in the lid so the tube can enter but nothing else can. Label the clean water bucket and never use it for anything else.

If you create a constant flow inlet (from a stream, for example), make sure you also create an overflow outlet by drilling a hole near the top of the drum and using tubing to direct the excess water to wherever you want it to go.

Cleaning

If you notice the flow getting low, the filter is probably clogged.

Pour a bucket of water in and swirl the upper layer of sand with your hand in a circular motion. Remove the dirty water created from swirling, then smooth out the sand on top. Repeat this "swirl and dump" procedure until the flow rate is restored.

It will take several days for the good bacteria to reform. Retest the water before drinking it.

You can clean the outlet, diffuser plate, lid, and outside without dismantling the whole thing. For everything but the outlet, use bleach or soap, then rinse with clean water.

Clean the outlet by back-flushing it—that is, forcing clean water through it in the opposite way from the one in.

FUELS

The fuels in this section are not technically forms of renewable energy (except biogas), but are useful for cooking and/or generating heat.

BEESWAX CANDLE

Beeswax is excellent for making candles. Unlike paraffin wax, it doesn't release carcinogens. In fact, it cleans the air. It also smells nice, with a faint honey scent. If you decide to keep bees, you will soon have a good supply of beeswax. If not, you can buy it.

Candles perform differently depending on the quality of the beeswax, the size and shape of the jar, and the thickness of the wick.

What You Need

- 250g beeswax.
- 1/4 cup coconut oil.
- 10 drops essential oil, such as lavender (optional).
- A small saucepan.
- A wooden spoon.
- 2 wide-mouth mason jars, sterilized with boiling water.
- A size # 6 cotton square-braided wick.
- A wooden skewer.
- Sticky tape.

Once you are done with something, wash it before the wax cools.

Directions

Put the beeswax in a glass jar, then place the glass jar in a small pot one quarter filled with water. Bring the water to a simmer to melt the wax.

While waiting, prepare your wick, but keep a close eye on the wax.

Cut the wick to the height of the jar plus 10cm (4in). Use a bit of sticky tape to stick one end of the wick to the center of the bottom of the other jar. Balance the skewer over the middle of the jar and tape the other end of the wick to it, so stands vertically in the center of the jar. You want it to be taut but not super tight. Tape the skewer in place on the jar.

Once the beeswax is melted, take it off the heat. Stir in the coconut oil and essential oils with the wooden spoon. Adding the coconut oil lowers the melting point of the beeswax, which will prevent it from burning straight down (tunneling). It will also cut the cost since it is cheaper than beeswax.

Pour the wax mixture around the wick. If you want a perfect top at the end, leave a little of the wax behind. Allow it to sit for at least two days at room temperature.

If the top cracks or dips from cooling, re-melt the mixture you left behind and top it up (optional). Trim the wick if needed.

A candle will never burn as wide as it does the first time it is lit. To prevent it tunneling, you want the first burn to be long enough so the entire top melts. This takes about one hour per every 2.5cm (1in) of the candle's diameter. For example, if your candle is 5cm (2in) in diameter, it will take approximately two hours for the top to melt.

If the burn can't reach the sides, your wick is too thin for the thickness of your candle. Get thicker wicks next time. If the wax drowns the flame, your wick is too thick. Get thinner wicks next time. You can estimate the size of wick you need here:

http://www.busybeecandlesupply.ca/wicking.html

FIREWOOD

Burning firewood releases a lot of emissions, which is not very environmentally friendly, but it can be an easy and free source of fuel depending on where you live.

Collection

Only collect dead, seasoned (aged) firewood. Cutting down trees is bad for the environment and doesn't make good firewood anyway. Collect tinder and kindling as well.

There are two types of wood to consider for firewood. Collect them both and keep them separated.

Soft wood is good for starting a fire. It is faster to dry and easier to light, and it burns quickly.

Hard wood is better for keeping a fire going. It is harder to ignite, takes longer to dry, and is harder to split, but it burns longer. The drier the wood, the better it is to burn. Some characteristics of dry wood are:

- It has cracks in the grain.
- It's lighter than wet wood.
- It will make a hollow sound when hit against another piece of wood. Wet wood will thud.

Splitting

Splitting wood makes it faster to dry and season and easier to store and use.

If it is very long, cut it to size first. Half a meter (20in) is a good length. Use an axe or maul to split it. A maul is heavier than an axe, with a wider head. It is better for splitting wood, especially for bigger pieces. A splitting wedge and sledgehammer may also come in handy. Sharpness helps, but is not critical. You are splitting the wood, not cutting or chopping it.

Make sure you are wearing gloves and safety glasses, and avoid splitting where there are nails or knots.

Set the piece of wood on another one and angle it up to cut it. This is safer and easier than the traditional method of standing it on its end.

Hit it on the end, along the grain. Chop down centrally, with your legs apart. Swing the axe up high (but controlled), then let it crash down into the wood as you slide your hand down the handle. Add some muscle behind it, but let the weight of the axe do most of the work.

If you do not hit the wood on the end it is less likely to split.

With larger/harder pieces that don't split all the way through, flip them over and hit them again from other side.

Storage

A common way to store wood is to stack it. It is best to do this once the wood is dry.

Stack it somewhere accessible and safe with a lot of airflow and sun. Consider sheltering it from the rain as well.

Using the log-cabin stacking technique will keep it stable. Any wood that is on the ground will get damp, so lay a few pallets down first if you want, then alternate vertical and horizontal pieces with each row as you build the stack up.

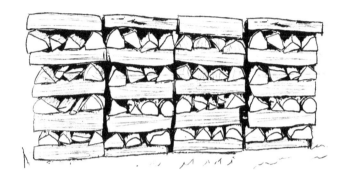

If you are stacking the wood between two things (trees or the walls of a shed, for example), you can face all the pieces in the same direction. This allows you to stack more in a smaller space. Stack pieces evenly to prevent the pile from toppling.

Never stack wood higher than your head, and always take the pieces you want to use from the top.

CHARCOAL

Charcoal is made from wood, but it burns more efficiently. Here are some pros and cons of charcoal vs wood.

Pros:

- It burns at a higher temperature.
- It burns longer.
- You can store more. It produces the same heat per pound, but charcoal is heavier.
- It doesn't absorb as much water as wood.
- It doesn't need to be seasoned.
- There is less smoke and soot when it burns.
- You don't have to cut or split it.
- It doesn't attract insects.

Cons:

- Coal ash is not good for the garden, unlike wood ash.
- It doesn't light as easily as dry wood.
- It leaves black marks when you handle it.
- There is no wood smell or taste.

What You Need

- A paint can with a lid
- Seasoned kindling
- A nail
- A hammer
- Matches

Directions

Clean the paint can very well or buy a new, empty one. Punch a hole in the lid with the nail and hammer. Put as much of the kindling into the can as you can, then put the lid on it.

Light a fire and put the can of wood in it. After a while, smoke will stream out of the hole. When the smoke stream becomes lighter, set a match to it. The match will catch on fire. If it blows it out, then the smoke is still too strong. Just try again later.

Once the smoke stream's fire goes out, try to light it again. If it does not ignite, the charcoal is ready. Remove the can from the fire and let it cool before retrieving the charcoal.

This project produces a small amount of charcoal, but it is easy to scale. Just use a metal drum with a lid instead of a paint can, punch a bigger hole, and make a bigger fire.

BIO-BRICKS

Making bio-bricks is a great way to recycle any old bio-trash such as newspapers, cardboard, dried foliage, napkins, sawdust, or paper plates. Don't use anything you wouldn't normally burn, like plastic.

What You Need

- 2 buckets
- Bio-trash
- Water
- A drill with a paint mixer bit and a normal drill bit
- A Tupperware container a little bigger than a brick
- A brick

Directions

Fill the bucket up with bio-trash and soak it all in water.

Use the drill with the paint mixer bit to mulch it all up. You could also do this with a cake mixer, by putting it bit by bit into an old food processor, or by cutting it up manually with large gardening shears. The smaller you can shred it, the better. If the pieces are too big it will fall apart when it dries.

Drill a bunch of holes in the bottom of the other bucket, then pour the mixture into it. This is to strain out the water. If you are doing this more than once, have large bowl (or another bucket) under the bucket with holes so you can reuse the water. Use the original bucket to push the mixture down into the bucket with holes to get more moisture out.

Drill holes in the bottom of your Tupperware container. This is your mold. Fill it up with bio-mulch and press as much water out as possible by placing a brick on top of it and then standing on it.

Any mold you want to make will work. The bigger it is, the longer it will take to dry, but the longer it will burn. The main thing is to press as much water out as possible.

Leave the bio-bricks in the sun to dry, then use them as you would firewood.

GEL FUEL

Gel fuel is an alcohol-based fuel that burns for a long time and produces no smoke or odor. It is safe to use indoors without ventilation, and unlike liquid fuels, it won't splash if it's knocked over.

Use gel fuel in fireplaces, grills, fire pits, to keep food warm, for camp cooking, etc.

Gel fuel is made with isopropyl alcohol and either soy wax or calcium acetate. It is cheap to buy as a finished product, or you can buy these two ingredients and make it yourself.

To get isopropyl alcohol, buy rubbing alcohol. The higher the concentration of isopropyl, the better. You need at least 70%, but 99% percent is better and not hard to find.

You can also buy soy wax or calcium acetate, or you can make calcium acetate.

Calcium Acetate

To make calcium acetate, mix four parts white vinegar with one part calcium carbonate (crushed chalk).

You need to evaporate one third to one half of the liquid. To do this, you can either:

- Leave it in the sun
- Put it in the oven at 90C (200F) for about three hours
- Cook it out on the stove over a low heat

It will smell like rotten eggs.

Gel Fuel

To make get fuel, mix nine parts isopropyl alcohol with one part of either calcium acetate or melted soy wax. Stir as you add the calcium acetate or melted soy wax to the isopropyl alcohol. Mix it well. You can add a little scented oil if you want.

If you bought pure dry calcium acetate, you need to mix it with water before mixing it with the isopropyl alcohol. Use a ratio of one part water to three parts calcium acetate.

If you're using soy wax, melt it the same way you would melt beeswax, by double-boiling it.

If your gel is not burning well or is too thick, add a little more isopropyl alcohol.

Storage

Pour your gel fuel into a non-flammable container, such as an empty soup can. Fill it three quarters of the way. Include a wick that goes to the bottom of the can.

Label it and put it in the fridge to cool overnight.

Once it's set, store it out of the heat and sunlight and out of reach of children and pets. It will store indefinitely.

BIOGAS

Biogas creates a clean energy source via the breakdown of bio-waste, like animal waste or food waste. It works best in hot climates.

This bio-gas "machine" creates two usable substances. The bio-gas is methane, which you can use for heat, cooking, etc. The second substance is a liquid fertilizer slurry. Dilute it with non-potable water at a ratio of 10 to 1 (10 parts water) to make a compost tea.

You will also get a buildup of non-liquid slurry on the bottom of the machine, which you can add to your garden once a year when you clean the machine out. If you have this biogas machine, you do not need a compost heap.

What You Need

- An IBC container.
- Foam spray (optional).
- Black paint.
- A drill.
- A 3in drill bit.
- A 2in drill bit.
- 2in uniseals.
- Plumber's silicone.
- 2 x 1.5m (5ft) of 2in PVC pipe.
- 1.5m (5ft) of 4in PVC pipe.
- 1m (3ft) 2in PVC pipe.
- A hacksaw.
- Sandpaper.
- A 2in to 1/2in PVC reducer.
- A 0.5in PVC ball valve.
- A 0.5in barbed hose connector.
- A gas hose. Clear plastic tubing is fine.
- 2in PVC T.
- 2 2in PVC elbows.

- A bucket.
- A large PVC bladder. Car or truck tires also work.
- Animal manure.
- Water.
- A funnel.

Directions

You can build/put this indoors or out. If you feed it correctly, there is not much odor. If you want to put it outdoors but you live in a cold area, insulate it with foam spray.

The IBC container is the digester. Paint it black so no light can enter it; otherwise, the gas will be CO_2. Drill three 3in holes in the top in three corners and put the uniseals in them.

Use the 4in pipe for the feeding pipe. Cut the bottom of it off at a 45-degree angle. Sand down the rough edges and insert it into one of the holes in the IBC container. Push it all the way to the bottom of the tank, with the angled end facing down.

Use a 2in pipe for the gas outlet. Drill a hole in the pipe 2cm (0.8in) below where the top of the IBC container will be once the pipe is inserted into the hole and pushed to the bottom. Sand the edges flush and put it in the hole on the same side of the IBC container as the feeding pipe.

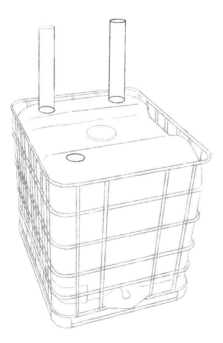

The other 2in pipe is the slurry pipe. It is where the compost tea will exit from. Make a 2in hole in the pipe so it sits approximately in middle of the tank.

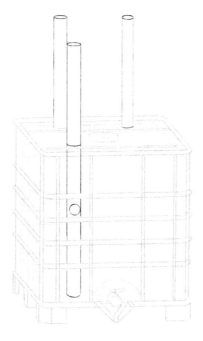

Put the 2in to 0.5in PVC reducer on top of gas outlet pipe.

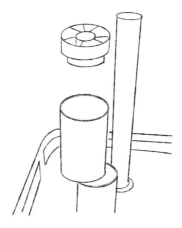

On top of that, put the 0.5in PVC ball valve, then a 2in elbow, then the 0.5in barbed hose connector. Do not glue them on, or you won't be able to clean them later.

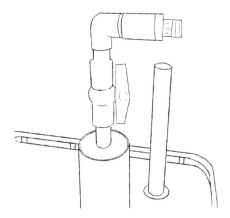

Cut 5cm (2in) off the top of the slurry pipe and attach the 2in T on top of it. Attach the piece of pipe you cut off into the side of the T, then the 2in elbow, then the 1m 2in PVC pipe, so it goes down into the bucket.

You must use a T as opposed to an elbow because the open part ensures there is no vapor lock. Make sure the 90 degree in the T is lower than the feeding pipe.

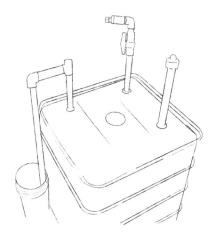

Connect the gas outlet to the PVC bladder with a gas hose and then connect the bladder to your stove.

Use the funnel on the feeding tube to add fresh animal manure (pig, cow, sheep, horse, etc.) and water. You want it to be the consistency of thick pea soup, which is approximately one part dung to ten parts water.

Wait one month. You can put a plug on top of the feeding tube to prevent odors if you want. Check whether your gas bladder is rising. If it does not start inflating within a couple of weeks, there is a leak. Once it's full, vent it to empty. This first lot of gas won't burn.

After one month, see if the gas will catch on fire. It may be hard to see in the daylight. If it doesn't catch, test it again once a week until it does. Don't add any food waste until you get the first flammable methane.

Once it lights, you can start adding a little food waste day by day. Blend it into a slurry using one part food waste to one part water. Any food waste is fine, including meat. Start slowly, adding no more than 1L of the slurry per day at the start. Eventually you can build up to one bucket a day, but never do more than that.

Pour a little vegetable or citronella oil in the pipes every couple of weeks.

Search for the following Facebook group to get community help: Solar CITIES Biogas

Here are some alternate design examples:

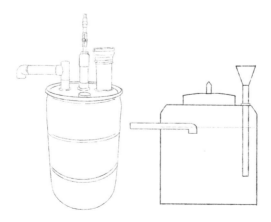

ALTERNATE ENERGY

In this section, you will get an overview of how to create electricity via renewable energy sources. There are also some projects for powering small appliances without using electricity from the grid.

ALTERNATIVE ENERGY

This is an overview of the various ways you can create electricity via renewable energy sources. It does not give plans for system setups, and unless you are very handy, you will need further research and/or professional help. What is does do is give you enough information to assess your choices, so you can decide which is the best option for where you live.

How Much Power Do You Need?

A quick way to determine this is to check your current electricity bill. If that is not possible (if you're in a camper, for example), you can figure it out by following this example. Note the power rating of every appliance that will draw from the system. For example:

- 2 x 15W 12VDC fluorescent lights
- 1 x 60W 12VDC water pump
- 1 x 50W 240VAC TV

Calculate the total DC and AC loads

AC Loads

- Television: 1 x 50W. Used 3 hours per day = 150Wh/day (watt hours per day).
- Total AC = 150Wh/day

DC Loads

- Lighting: 2 x 15W DC Lights. Each used 4 hours per day = 120Wh/day
- Pump: 1 x 60W DC Pump. Used 1/2 hour per day = 30Wh/day
- Total DC = 150Wh/day

Divided by 0.85 to allow for inverter efficiency (85%)

- 150/0.85 = 130Wh per day (rounded up)

DC + AC = Total load needed

- 150 + 130 = 280Wh per day

In this example, you would need to create at least 280 Wh/day.

Solar

Solar electricity is the most accessible renewable energy source for most people.

There are two types of solar energy. Thermal solar energy is for things like heating and cooking. Photovoltaic (PV) solar power is what creates electricity. Patio solar lights are a good example of PV solar energy.

You need to assess the weather in your location to see if it gets enough sun to be worth installing a PV solar power system. Consider:

- Where you live on the planet
- The amount of sunlight you get throughout the year
- The placement of the solar panels

Here is the formula for power generation from solar panels. It is an approximation only.

(Solar panels' combined maximum watt output) x (average hours of sunlight per day) x (0.5) = daily sun power generated

0.5 = estimated power loss due to system inefficiencies and conversion factors.

To convert watts to kilowatts, divide watts by 1,000.

Hydro

Hydro-electricity uses flowing water to generate electricity. It is the most reliable form of renewable energy.

Under the right conditions, it will generate electricity non-stop, and a micro hydro generator in a good location can produce up to 25 kWh/day, which is enough to power an average household. It is also cheaper than installing solar panels.

Unfortunately, you need flowing water running through your property to access it, which isn't common, and even if you do have it, it may be regulated by the government.

If you have year-round flowing water on your property, this is something you should look into. Ideally you want high pressure and water volume, though issues like rising and falling water or light flow can be overcome.

DIY installation of a micro hydro generator is relatively easy compared to solar.

If you don't want to pay for a hydro generator, consider building a water wheel to harness the flow of water for mechanical means.

Wind

Windmills are for mechanical processes, like pumping water. To create electricity, you need a wind turbine. If you get a constant wind of at least 5km/h (3mph), this may be worth looking into, but it is the least reliable of the three options.

If you live in a coastal area, on an open plain, or on a hilltop, a small wind turbine about 180cm (6ft) in diameter can be a good supplemental power source and is not too difficult to DIY from recycled materials.

Combining Renewable Energy Sources

Combining renewable energy sources is a good way to build up a decent amount of power, and unless you have access to running water source, a wind-and-solar combination is your best bet. If you want detailed instructions, visit:

https://gumroad.com/a/501331059

SODA BOTTLE LIGHT BULB

This is a way to light a dark room, such as a shed with no lighting, on a sunny day. It doesn't create or use electricity and is more like an alternative to installing a skylight.

With this method, a water bottle is filled with water and installed into the roof. The water refracts sunlight and spreads it out into the room.

The design below is for installations on a metal sheet-roof shed, but it is easy to adapt to other types of roofs.

What You Need

- A clear water bottle with lid and no scratches or leaks
- Water
- Bleach
- A piece of sheet metal roofing
- Tin shears
- Sealant

Directions

Remove any labels from the bottle and clean it. You want it as clear as possible.

Cut a hole in a square piece of sheet metal roofing the same diameter as the bottle. Put the bottle in it, so that the bottom of the bottle will face down into the roof. Use sealant around it so the roof won't leak when it rains.

Fill the bottle with filtered water and two capfuls of bleach. This ensures no bacteria will grow inside it.

Cut a hole in the roof the same size as the bottle.

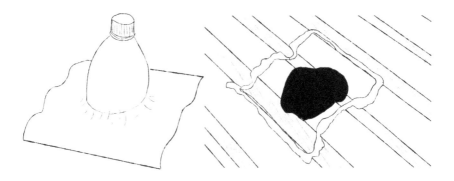

Insert the bottle into the sheet metal roofing, and seal it well to prevent leaking. After a while, the water will evaporate. Top it up with water and bleach.

PEDAL POWER

Pedal power is when you rig a bike to convert your physical movement into mechanical energy. It is like a hand tool, but uses your legs.

There are many guided projects online for a wide variety of pedal-powered machines, and you can adapt them to almost any small appliance, such as a blender, water pump, oil press, or washing machine.

There are three basic methods for doing this.

The first is to use a friction wheel against the bike wheel to redirect the turning energy. Use a stand to suspend the wheel in the air.

Here is a couple of basic designs for a pedal-powered blender using this method. Find plans for this and other projects at:

http://mayapedal.org/machines.en.

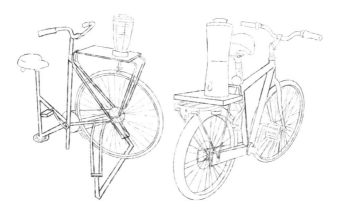

The other method is to use the direct friction from the spinning wheel with whatever you want to spin, like this pedal-powered washing machine. Find detailed plans here:

https://www.askaprepper.com/make-semi-automatic-off-grid-washing-machine-no-electricity

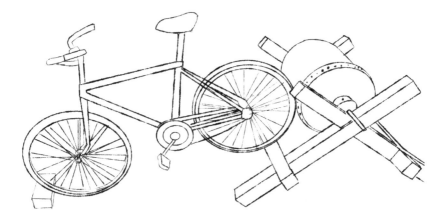

A third option is to use the chain, like in this pedal-powered compost turner.

No matter which method you use for your own projects, it is best to keep it as simple as possible. Fewer moving parts mean less can go wrong.

PEDAL GENERATOR

A pedal generator takes pedal power up a notch, and converts your physical energy into electricity. You can use it to charge a battery (like your phone) or use the electricity directly to power low-wattage electrical appliances (like a radio). This is actually an inefficient use of energy, but is a good project for learning.

As you pedal, it rotates the motor which acts as a generator and converts your mechanical energy into electrical energy. The diode ensures the current flows in the right direction and the battery stores the current. Finally, an inverter converts the DC power from the battery to AC power, so you can use it for small appliances.

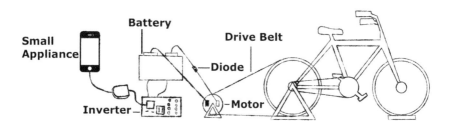

What You Need

- An old bike.
- Wood.
- A wrench.
- A V-belt, also known as a drive belt. Get one from an auto parts shop.
- A saw.
- A diode.
- A hammer and nails (or screws and screwdriver).
- A car battery or one of similar size.
- An inverter.
- A tape measure.
- Electrical Wire.

- A 12V Motor (or higher) with a mounting bracket.

Remove the back tire of the bike.

Build a stand from wood to raise the back wheel of the bike 15cm (6in) off the ground. It must be stable enough to keep the bike elevated and secure as you pedal.

Attach the V-belt to the rear wheel of bike, and then to the motor shaft. It must be tight. Ensure that the tire and motor shaft spin together, with no slipping.

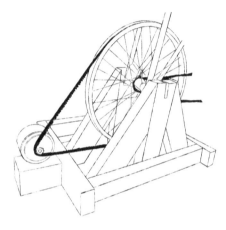

Connect the positive motor lead to the anode side of the diode with electrical wire. Continue the circuit by connecting the cathode side of the diode to the positive terminal of the battery.

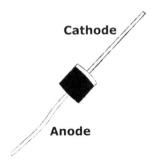

Connect the negative motor lead to the negative battery terminal.

+ve Motor lead — anode - Diode - cathode — +ve Battery terminal

-ve Motor lead — -ve Battery terminal

Connect the battery leads to the inverter, positive to positive and negative to negative. You can use an adapter to do this, or hard-wire it by either soldering the leads or twisting the wires together and covering them with electrical tape.

Use a multimeter if you want to check how much voltage you create while peddling.

Related Chapters:

- Pedal Power

SOLAR POWERED APPLIANCES

This small solar-powered setup will power almost any small household appliance. If you want more power, just scale everything up.

Calculations are for small-scale systems only. For larger ones, such as for a house, seek professional help, as there are many other factors to consider.

When selecting gear, always choose something with more capacity than you think you will need.

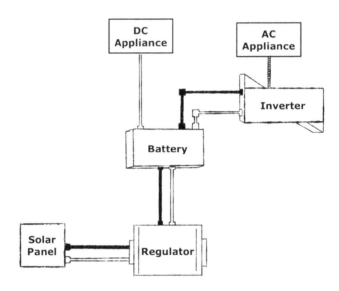

Solar Panel

The solar panel converts sunlight into DC electricity to be stored in the battery.

Every quality solar output will have a rated output which is for one peak hour of sun. Hours of peak sun received depend on your location, season, and weather.

You can wire multiple solar panels together to increase voltage (in series) or current (parallel).

To calculate the required solar panel input, use the following formula:

*(Wh needed/average peak sun hours) * 1.4 = Required solar panel output*

To discover wattage per hour needed, see the Renewable Energy chapter. The 1.4 is a constant and simplifies the calculations for basic solar systems.

Here is an example:

- (189Wh / 6h) * 1.4 = 45W

You would need a solar panel with a rated output of at least 45W.

Regulator

The regulator ensures the battery is charged with the correct voltage. It also prevents back-feeding from the battery to the solar panel at night.

You may lose some energy with a regulator, but it will prevent the battery from getting damaged by overcharging.

Ensure your solar regulator has a low-voltage disconnect and can handle at least 125% of your solar panel's rated short-circuit. In high-temperature regions, it is best to have at least 150%. In addition, consider whether you will want to add more solar panels in the future.

Deep-Cycle Solar Battery

The battery stores the electricity. A deep-cycle solar battery is designed to discharge over a long period and be recharged thousands of times. Batteries are generally 90% efficient.

You want a big enough battery to supply your total power usage without being discharged more than 70%, and one with a backup capacity of four days. This allows for low-sunlight days.

You can join batteries together to achieve this.

Here is the formula:

(Total Wh * 4 days / Battery Voltage) / 0.7 * 1.1 = Ah required

The 0.7 is the 70% discharge. The 1.1 is for the 90% efficiency of the battery.

If using the example of 280Wh per day and a 12V battery:

- Ah Required = (280Wh * 4 / 12V) / 0.7 * 1.1 = 122Ah. (rounded up)

*Dividing by 12V converts the Wh to Ah.

Inverter

The inverter converts DC current to AC. Most small appliances run on AC.

The inverter must be capable of supplying the maximum anticipated combined AC load. More is better. It must also have a suitable surge rating to cope with the start-up of appliances.

There are two outputs you can choose from when purchasing your inverter. Pure sine wave is almost identical to power from the grid. It is the preferred output, but is more expensive. Modified sine wave is cheaper, but not all appliances will work optimally with it.

SIMPLE SOLAR SETUP FOR DC APPLIANCES

This is also a simpler system for powering 12V DC appliances. Examples of 12V DC appliances include DC fans, LED lights, and USB devices.

This design replaces the regulator and battery in the previous project with a step-down converter. The step-down converter modulates power from the solar panel to suit your appliance's voltage needs.

A step-down converter can be 12V, 24V, or 36V. It will only produce the required voltage when the entire solar panel is in full sunshine.

Ensure your step-down converter has over-current/voltage/temperature protection.

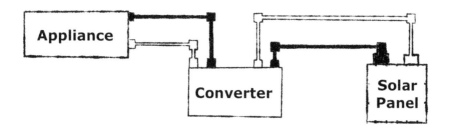

Solar Panel

The solar panel must exceed the amount of power required by the appliance so it can cover conversion loss. For example, if you have a 20-watt fan, use a 50-watt solar panel.

It also needs an open circuit voltage that is less than the maximum input voltage of the converter. You can join solar panels in parallel if needed, though converters can typically only handle 40 volts anyway.

Converter

The converter must be able to handle the voltage output of the solar panels.

Solar Panel Wattage / Open Circuit voltage of Panel = Amp output of solar panel = Max input in amps that the converter can handle

Troubleshooting

If there is no damage to your components and your appliance isn't working, it means there is not enough power. A common mistake is that your solar panel is not large enough to fulfill the startup surge requirement of your appliance.

HEATING AND COOLING

Heating and/or cooling your home can use a lot of energy so it is worthwhile to adopt passive methods such as:

- Cooking indoors or outdoors, depending on where you want the heat.
- Insulation.
- Thermal or cooling curtains.
- Vegetation.
- Weather stripping.

You can purchase thermal and/or cooling curtains or make them yourself. To make thermal curtains, get some heavy blankets and sew in some extra insulation. Any reflective sheet, such as a mylar survival blanket, makes a good cooling curtain.

Although not 100% passive, low-powered fans use much less energy than air-conditioning. Attach a water mister to a fan for increased cooling. You can also rig it up so it runs on solar power.

Geo-thermal energy uses the consistent temperature of the earth to heat or cool your home. It is reliable and is available almost everywhere, but costly to install. If you are very handy you can get free plans off the internet and rig it up yourself. A DIY geothermal system is not an easy project, so do your research.

PORTABLE SODA-CAN SOLAR HEATER

This portable heater harnesses thermal energy from the sun. It absorbs and contains the heat, then fans it out to where you want it. It can produce 65C+ (150F+) air on a cool (10C/50F), sunny day.

What You Need

- 24 undamaged aluminum cans. Scratches are okay but no dents, cracks, or tears.
- Weather-proof matte black paint rated for use with high heat.
- Plexiglass 45cm (18in) x 60cm (24in).
- Cardboard.
- 0.25in plywood.
- "1 x 4" wood. 2.5m (8ft) long is enough.
- A small solar fan.
- A 3in duct vent.
- Caulk or glue rated for high-heat use.
- A drill.
- Screws.

Directions

Sand the surfaces of the cans so the paint will stick to them better. Break off the tabs and punch three holes in the bottom of every can. Paint them black.

Glue four cans into a tower, with the openings of the cans facing down. Create six of these towers.

Make a cardboard frame with the following dimensions:

- 45cm (18in) x 60cm (24in).
- 8cm (3in) wide. This is the width of each side of the frame.
- 2cm (0.75 inch) thick. Glue several pieces together to achieve this.

The cardboard frame is insulation. You can use different insulation (foam board, for example) if you want.

Create six 5cm (2in) holes along the center of the bottom side of the cardboard frame. This allows air to enter the heater. Each hole must line up with the center of each can.

Glue the frame together and paint it black.

Next, you will make the wooden frame. Measure the sides using the cardboard frame as a template, but leave a gap at the bottom to allow air to get through. You can make a little support shelf for the cardboard frame to sit on. Use the plywood for backing and paint the whole thing black.

Install the solar-powered fan near the top, so it's facing the back of the frame. Stick the solar panel on the front of the frame. Depending on your solar fan, you may need to take it apart to do this. Surround the fan with the duct vent to channel the air. Put the cans inside and seal them in with the plexiglass. Seal all gaps with the caulk or glue. It needs to be airtight.

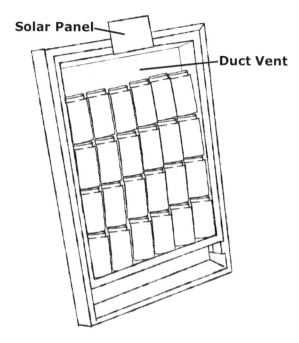

Face the cans toward the sun. The cool air will enter from the bottom and warm up as it passes through the cans, then get forced out the top with the fan.

There are plans on the internet for larger versions of this that you can attach to the side of your home.

SOLAR HOT WATER SHOWER

This uses the thermal energy of the sun to create enough hot water for a shower. It needs the sun to warm up, but once it's heated, it will stay warm without the sun for hours. It also incorporates a cold water source so you have temperature control. It makes a great outdoor shower, but you could also run it inside.

What You Need

- 30+m (100+ft) of 0.5in black poly irrigation tubing. The more tubing you have, the more hot water you get.
- 2 black poly garden-hose-style connectors.
- 15+m (50+m) green garden hose. This is for the cold-water line.
- 2 Y splitters.
- A small spot sprinkler. This will be the shower head.
- A sheet of plywood.
- Twine.

Directions

Coil the black tubing on the plywood and tie it down. This makes it easier to move the tubing around. Place it somewhere that gets a lot of sun, like on a tin roof, or angle it towards the sun.

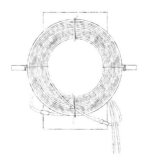

Attach the Y splitter to your tap. Connect the black tubing to one end, and the garden hose to the other.

Attach both hoses and the shower head to the other Y splitter.

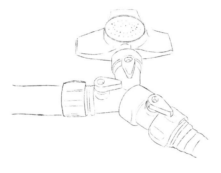

Fix your shower head where you want it.

Run water through the black pipe and wait for it to heat up. This will take about 20 minutes on a sunny day.

When you want to use it, turn on the cold water first, then add hot water until you get your desired temperature. This prevents you from wasting hot water or getting burned.

Solar Batch Water Heater

An upgraded version of this project is to install a solar batch water heater. This will store more water and you can plumb it into your domestic water line.

You can buy one or build it yourself, though you may need help if you want to plumb it in.

There are free, detailed instructions on the internet, but the basic idea is to get an old hot water tank and paint it black. Place the tank inside a wooden box that has a reflective inside and plumb it to the hot-water mains.

If you don't want to plumb it in, hook it up to a hose and use gravity to create the water pressure.

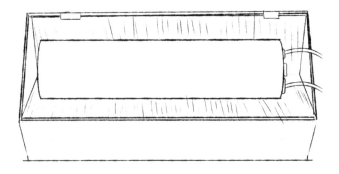

HEALTH AND HYGIENE

Companies put a lot of chemicals into health and hygiene products, some of which are linked to various forms of cancer. There are many natural alternatives, but they cost quite a bit more in comparison.

Making your own means you know exactly what is going in it and is considerably cheaper.

The projects in this section cover how to make:

- Soap
- Various natural health- and hygiene-related sundries
- Natural pest-control measures

All the "recipes" in this section have a focus on minimizing ingredients, often using things you can procure through other projects in this book, like beeswax.

The rest of this introduction covers some health- and hygiene-related subjects that aren't projects but are useful to know.

Coconut Oil

Pure cold-pressed coconut oil is a versatile product that needs no extra preparation. You can use it straight out of the bottle as a(n):

- Antibacterial. Apply it to minor cuts and scrapes.
- Lip balm.
- Moisturizer.
- Mouthwash/oil pull.
- Balm to soothe skin irritations.

Vinegar

Mixing any pure vinegar (white or apple cider vinegar, for instance) in a one to one ratio with water makes a great all-purpose surface cleaner. Add essential oils if you want to scent it.

You can also use it in conjunction with baking soda to unclog drains. Put half a cup of baking soda down the drain, then follow it with one cup of vinegar.

Soap Plants

You can use plants that contain saponin as soap. Native varieties exist all over the world, but each is used differently. Research soap plants for your specific area.

Toilet

Unless you are living somewhere with no plumbing, it is better to stick with what you have. You can replace your toilet with a water-saving one if you don't have one already.

If you don't have plumbing or want to do your business outdoors for whatever reason, you have a couple of options.

A drop toilet is a deep hole in the ground. Cover it with a portable outhouse on sled runners so it is easy to move once the hole is full.

Keep it dry and ventilated. Include a bucket of dirt so you can throw some down the hole after each use, and have a hand-washing station nearby.

Placement of your drop toilet is important. Put it:

- At least 15m (50ft) from any water sources.
- Downwind of your home or anywhere you don't want the smell.
- Within 30m (100ft) of your home for convenience.

A permanent composting toilet is almost the same as a drop toilet, but it composts your waste so you don't need to move it.

Using a septic tank is the best long-term solution. The waste will go from your toilet to the tank, then into a drain field. Make it a downhill system to take advantage of gravity, otherwise it needs to pressure based which is more expensive to install.

Double Boiling

Double-boiling is used for safety when melting highly flammable substances. You can buy a double-boiler or use a jar and a pot.

Put what you want to melt in a sterile glass jar, then place the glass jar in a small pot one quarter filled with water. Bring the water to a simmer and stir whatever is in the jar while it melts.

Related Chapters:

- Vinegar
- Apple Cider Vinegar

CASTILE SOAP

Castile soap is an all-natural, vegetable-based soap. You can use it with a water dilution (to whatever strength you want) or as a base to make other hygiene products.

You can use if for:

- Clothes soap.
- Deodorant (two parts sea salt to one part Castile soap with water in a spray bottle).
- Dish soap.
- Hand, face, and body wash.
- Household cleaner (surfaces, floors, bathroom, windows, etc.).
- Insecticide on plants (use a very mild dilution).
- Makeup remover (with equal parts witch hazel and a carrier oil).
- Mouthwash (with peppermint essential oil).
- Pet soap (do not add any essential oils or make it with avocado oil).
- Shaving cream.
- Toothpaste (with peppermint essential oil).

There are specific projects on how to make bar and liquid versions of Castile soap. This chapter covers some general information.

To make Castile soap, you need vegetable oil and lye. Traditionally, it is made from olive oil, but other oils you can use include coconut, castor, hemp, avocado, walnut, and almond. The oil is needed to lather, moisturize, and cleanse.

Different oils are good for different things. For example, olive oil is gentle on the skin, whereas coconut oil lathers more and is better for non-body cleaning, such as around the home or of clothes.

Using different types of oils also requires different measurements. Use this online calculator:

http://soapcalc.net/calc/soapcalcwp.asp

The other ingredient, lye, is caustic. When making or handling lye, safety is paramount. Here are some guidelines:

- Ensure there is adequate ventilation.
- Keep pets and kids and a safe distance.
- Keep an eyewash solution (sterile saline or distilled water) handy.
- Store it securely.
- To avoid chemical reactions, do not use most metals. Stainless steel is okay.
- Wear rubber gloves, a rubber apron, goggles, and long-sleeved clothing.
- Work slowly and gently to prevent splash-back.
- Work near a reliable supply of cool, running water.

If you get lye on your skin, rinse it under lots of cool running water for at least 10 minutes. After rinsing, if there is still a burn, put a little vinegar to neutralize it. Do not go straight to vinegar—always rinse first. Seek medical attention.

If it gets in your eye, rinse it out with eyewash solution for at least 10 minutes and seek medical attention.

Additions

There are many things you can add to your soaps to customize them. They work best in soap bars, but some can go in liquid soap too.

To add color, use herbal tea. Infuse it in the water before adding the lye. Alternatively, use herb powder. Clays are also good for coloring and have the added benefits of making the soap good for shaving and skin tone.

Some herbal powders, flowers, etc. will turn brown when added, but calendula, lavender, and nettle leaf work well. Research the ones you want to use to see if they hold color. To add exfoliators, grind one or more of the following. Grind it small, but not to a fine powder. Some of these, like coffee, will also add color also.

- Coffee
- Flower petals
- Leaves
- Oatmeal
- Salt
- Seeds
- White pumice

For scenting, you can use either essential oils or fragrance oils. Fragrance oils are synthetic, but they are cheaper, come in more varieties than essential oils, and are still skin-safe.

In most cases, stir in the additions to your soap at the end of the cooking stage. Right after turning the heat off but before putting it in the mold. To create marbled effects, add powders after the essential oils, since you don't want to stir them in as much.

Simple DIY Soap

If you don't want to mess with lye, but still want to make your own soap, you can purchase blocks of Castile soap that have been specially processed and melt them down. Make sure the soap is pure with no added chemicals.

Buy the blocks in bulk from an online distributor to save money. Melt the soap down over a low heat, add whatever you want, then pour it into your molds.

Related Chapters:

- Lye

LYE

When you want to make your own Castile soap, you need lye. Here you will learn how to make it, but if you can, it is better to buy it. Lye that is manufactured commercially is higher-quality and safer to use than anything you can make in your back yard.

There are two types of lye. Potassium hydroxide lye (KOH) is for making liquid soap. The other type of lye is sodium hydroxide (NaOH), and is for making bar soap.

This recipe makes potassium hydroxide lye (KOH), so you can only use it to make liquid soap.

If you're making your own lye, safety is paramount. Here are the guidelines again:

- Ensure there is adequate ventilation.
- Keep pets and kids and a safe distance.
- Keep eyewash solution (sterile saline or distilled water) handy.
- Store it securely
- To avoid chemical reactions, do not use most metals. Stainless steel is okay.
- Wear rubber gloves, a rubber apron, goggles, and long-sleeved clothing.
- Work slowly and gently to prevent splash-back.
- Work near a reliable supply of cool running water.

If you get lye on your skin, rinse it under lots of cool running water for at least 10 minutes. After rinsing, if there is still a burn, put a little vinegar to neutralize it. Do not go straight to vinegar — always rinse first. Seek medical attention.

If it gets in your eye, rinse it out with eyewash solution for at least 10 minutes and seek medical attention.

What You Need

- White hardwood ashes. Collect them every time you have a fire and keep it dry in an airtight container until you are ready to use them. It must be hardwood, because softwoods don't have enough potassium.
- 1 large lye-safe bucket. Use heavy-duty plastic, stainless steel, or wood. Metal will react with the lye, but stainless steel is okay.
- 2+ lye-safe buckets that are smaller than the first one.
- Soft water with no added chemicals. Use rain or filtered water.
- Bricks.
- Stones.
- Straw or grass clippings.
- A drill.
- Safety gear (gloves, safety glasses, long clothes, etc.).

One of the buckets is the lye bucket, and the other two smaller ones are collection buckets. The lye bucket can be as big as you want, but you will also need more and/or bigger collection buckets.

Directions

Find somewhere to make the lye where it will not be disturbed by the weather or anything else.

Drill several small holes in bottom of the lye bucket.

Make a stand for the lye bucket out of the bricks. You need it to be high enough so you can place a collection bucket underneath it. Make sure it is sturdy.

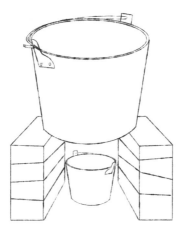

Put a 2.5cm (1in) layer of stones on bottom of the lye bucket. It needs to completely cover the bottom. Use stones bigger than the holes you made. On top of the stones, make a 7.5cm (3in) layer of straw or grass. These layers are a filter so the lye can drain out but the ashes will not.

Fill the rest of the bucket with cool ashes. Add the soft water until it is a slurry. You want it wet, but not too watery. Once the ashes begin to float and you can see the waterline, there's enough water.

Covering the lye bucket with a lid is optional. It will keep contaminants out, but make sure it is easy to take off to prevent any mishaps later.

After a few hours, lye water will start to come out. When the collection bucket is about 75% full, replace it with another bucket.

Test the concentration of the lye with one of the following methods:

- Dip a chicken feather in it. If the feather dissolves, the lye is good.
- Put a potato in it. If it floats, the lye is good.
- Use a pH tester. You want it pH 13+.

If the lye does not pass your test, carefully pour it back into the lye bucket. Repeat this test-and-return procedure until it is ready.

LIQUID CASTILE SOAP

The good thing about making liquid Castile soap, as opposed to a bar of it, is that you can add things after you've made it.

Make it pure, then mix different soap bases or add various substances depending on what you need it for.

This recipe makes about 2kg (4.5lbs) of undiluted Castile liquid soap. Don't forget the safety precautions as outlined in the *Castile Soap* chapter.

What You Need

- 185g (6.52 oz) of KOH lye. This is not the same as NaOH lye!
- 907g (32 oz) of pure olive oil.
- 327g (11.55 oz) of filtered water.
- 227g (8 oz) of glycerin.
- A stainless-steel bowl.
- A glass bowl.
- A wooden spoon.
- An immersion blender.
- A clean bottle.
- A funnel.
- A strainer (optional).

If you want to use a different type of oil, make sure you adjust the measurements of the other ingredients to suit:

http://soapcalc.net/calc/soapcalcwp.asp

Directions

Heat the olive oil over a low heat. Use a slow cooker if you have one. Mix the glycerin and water together, then add the KOH lye to it. Add the lye slowly, and make sure you have plenty of ventilation.

It is very important that you add the KOH to the water, and not the other way around.

Stir the liquid mixture until the KOH dissolves and the water clears up. Slowly add this mixture to the warm olive oil and stir it together while still on a low heat. Blend it until it goes thick, then creamy, then grainy, and then creamy again. If it gets too thick to blend, switch to a wooden spoon.

When you see translucent streaks, you can switch to a wooden spoon (if you haven't already) and continue to stir it for 10 minutes. Let it sit for three hours, stirring it every half an hour.

After three hours, test it. Dissolve a golf-ball-sized blob in half a cup of hot water, stirring it gently. Allow it to cool. If it is translucent (it won't be colorless) it is ready to dilute. If it is cloudy or has floating oil, cook it for another 30 minutes then check it again.

Even most bought Castile soaps are undiluted. Decide on the dilution level depending on what you want to use it for. The soap does not need to be thick to work well. It is liquid soap, not a shower gel. Here is a good resource for dilution levels:

https://www.drbronner.com/all-one-blog/2017/06/dilutions-cheat-sheet-dr-bronners-pure-castile-soap/

To dilute the soap, pour the required amount of distilled water into the soap paste (which is still on a low heat) and stir it every 30 minutes or so until it is completely dissolved.

If you want to make several different dilutions, remove the amount of soap paste you don't want to use yet. You can store it in any clean plastic tub and use it later.

The dissolving process can take up to eight hours. If after eight hours there are still chunks of soap, add more water.

Once done, turn off the heat and make any additions that you want, as outlined in the Castile Soap chapter. Transfer it to a clean bottle with a funnel. You can strain it if you want.

Related Chapters:

- Castile Soap
- Lye

COLD PRESSED CASTILE SOAP

This project creates the bar version of Castile soap. Bar soap isn't as versatile as liquid soap and it takes longer to make, but you can use a wider variety of additions, and the resulting products are great as homemade gifts or to sell.

Don't forget the safety precautions outlined in the *Castile Soap* chapter.

What You Need

- 1L (35 oz) of pure olive oil.
- 130g (4.5 oz) of sodium hydroxide lye (NaOH).
- 260g (9.2 oz) of filtered water.
- Sodium lactate (optional).
- Mold(s). Can be silicone, plastic, or glass containers. Line plastic and glass molds with oil or greaseproof paper to prevent the soap from sticking.
- A stainless-steel bowl.
- A glass bowl.
- An immersion blender.
- A wooden spoon.

Olive oil takes longer to harden in the mold than hard oils like coconut. Adding sodium lactate will speed up the process. Use one teaspoon of sodium lactate per pound of oil in the recipe.

If you want to use a different type of oil, make sure you adjust the measurements of the other ingredients to suit:

http://soapcalc.net/calc/soapcalcwp.asp

Directions

Put the filtered water in the stainless-steel bowl, then carefully add the lye in. Stir the mixture until the lye dissolves and it is clear. It is very important that you add the lye to the water, and not the other way around.

Pour the olive oil into the glass bowl, then pour the lye water into the oil. Blend it until it has a mayonnaise-like texture, then use the wooden spoon to stir in any additions, as outlined in the Castile Soap chapter.

Pour the mixture it into the mold (or multiple molds). Tap the mold on a hard surface to get rid of any bubbles then, if you want, smooth out the top with the back of the spoon.

Cover the mold with a lid or cardboard and wrap it in a blanket to insulate it. After 24 hours, check if the soap has set enough to remove from the mold. If not, wait longer.

Once it's set, remove the soap from the mold and cut it into bars if you need to. Smooth the edges if you want.

Cure the soap for at least one week (preferably a month or longer) by storing it in a cool, dry, and well-ventilated area. The longer you cure it, the harder it will get, and the longer it will last when you use it. Turn it once a day for a week, then once a week for a month, then once a month for however long you want to cure it.

Related Chapters:

- Castile Soap
- Lye

DEODORANT

The odor your body gives off is related to what you put into your body and the bacteria that grows on it. Besides using basic hygiene, to minimize your body odor naturally, you can eat clean. Consume less red meat and fewer spicy foods, and/or shave your pubic hair, which is where the bacteria likes to grow.

If you do not want to do those things, or if they are not enough, you can apply deodorant or antiperspirant. Unfortunately, most commercial deodorants and antiperspirants are filled with toxic chemicals to give them a "pleasant" smell. If you want to use them to mask your smell (like perfume), apply them on your clothes as opposed to your skin.

Here is a basic all-natural recipe you can make at home that will combat one of the root causes of body odor, the bacteria.

What You Need

- 1/8 cup of baking soda.
- 1/8 cup of corn starch.
- 2.5 tablespoons of pure coconut oil.
- 2 tablespoons of liquid Castile soap.
- 4 drops of tea tree oil.
- A glass jar.
- A pot.
- A container for storage.

Directions

Double-boil the soap and coconut oil together. See the introductory Health and Hygiene chapter for instructions on double boiling.

Stir in all the other ingredients. Pour the mixture in any clean container and let it cool before using it.

SUNSCREEN

The best way to prevent getting sunburned is to stick to the shade, put on a hat, and cover your skin with clothing. However, when you want to swim at the beach or feel the warmth on your skin, sunscreen is the way to go.

At around SPF 20, this homemade sunscreen is not as strong as most commercial ones, and it isn't waterproof, but it also doesn't have all the bad chemicals in it.

For safety, wear a face mask when handling zinc oxide powder.

As a side note, this same recipe without the zinc oxide makes a great skin lotion, but loses most of its sun protection qualities.

What You Need

- 0.5 cup of olive oil.
- 0.25 cup of coconut oil.
- 0.25 cup of beeswax. This is variable. The more you use, the thicker it will be.
- 2 tablespoons of zinc oxide powder. Get the kind that is non-nano and not coated.
- Essential oils for fragrance (optional).
- 1 teaspoon of Vitamin E oil (optional). This is for your skin and prolongs the shelf life of the sunscreen.
- A glass jar, sterilized with boiling water.
- A pot.
- Container for storage.

Directions

Double-boil the olive oil, coconut oil, and beeswax in a large glass jar. See the introductory Health and Hygiene chapter for instructions on double-boiling.

Once you remove the mixture from the heat, stir in the zinc oxide powder, essential oil, and vitamin E oil. Mix them in well, then pour the mixture into your container. Stir it every now and again as it cools. Store it in a cool, dry place or in the fridge.

LIP BALM

This lip balm works just as well, if not better, than any commercial lip balm, and it is 100% natural. The beeswax and coconut oil give it a mild but nice flavor and scent.

As a side note, this same recipe without the beeswax makes a great hand cream.

What You Need

- 2 parts coconut oil.
- 1 part beeswax. This is variable. More beeswax will make a thicker salve.
- A few drops of vitamin E.
- A glass jar, sterilized with boiling water.
- A small pot.
- Storage containers.

Directions

Double-boil all the ingredients together. See the introductory Health and Hygiene chapter for instructions on double-boiling. Stir the mixture until everything is melted together, then turn off the heat. Pour the mixture into your storage containers, and leave it to cool and set.

CAYENNE SALVE

This is like a homemade tiger balm that uses cayenne powder as the main active ingredient. Use it to relieve aches and pains.

You can swap out some or all the cayenne pepper to make other salves. For instance, you can use peppermint for a cooling salve to soothe dry skin.

What You Need

- 0.5 cup olive oil
- 15 grams cayenne powder
- 0.5 cup beeswax
- Cheesecloth
- A glass jar
- A small pot
- A storage container

Directions

Put the cayenne powder and olive oil in the glass jar and double-boil it. See the introductory Health and Hygiene chapter for instructions on double-boiling.

When the oil is hot, turn off the heat and let it sit for 20 minutes. Warm it up again, then let it sit again. Do this warm-and-cool cycle four times to infuse the oil with the cayenne powder. Once it is infused, strain the oil through the cheesecloth.

Put the beeswax in the jar and double-boil it until it is melted. See the introductory Health and Hygiene chapter for instructions on double-boiling.

Pour and stir the cayenne-infused oil into the beeswax until they are completely combined.

Pour the mixture into your storage container and let it cool and set.

To use the salve, apply it topically. Don't touch any sensitive areas, such as your eyes, armpits, or lips, after handling it. If your skin is very sensitive, you may get blisters. In that case, stop using it completely.

Making Cayenne Pepper

If you grow hot chilies in your garden, you can make your own cayenne pepper. Wear latex gloves while doing it.

Cut the stalks off your chili peppers and discard them. Dry the chilies using any drying method you want. Once they're dry, crush the chilies into a powder. You can use a grinder or do it by hand with a mortar and pestle. Crush them up as much as you can, and then sift the powder out. Put the part that isn't powder back into whatever you are using and crush it more. Do this several times until it is all powder. You may need to throw a little out.

Store it in an airtight container at room temperature.

HERBAL MEDICINE

Herbal medicine is using medicinal plants for their health benefits. There are many different medicinal plants, and even more different types of herbal remedies. Going into detail on the subject is far out of the scope of this book.

Instead, this chapter will give you an overview of the two easiest ways to extract the medicinal value from a plant: with water (tea) or alcohol (tinctures). Different plants require different extraction methods. Research the herb you want to use for the best way to do it.

This does not replace professional medical advice under any circumstances.

Tea

You can make hot or cold teas. To make it hot, put the plant part(s) in a pot or cup. Boil water and allow it to cool until it isn't boiling any more before pouring it into the pot or cup. Allow it to sit until the plant sinks and cools enough to drink.

For a cold tea, put the plant part(s) into a container and fill it with filtered water. Put it in fridge for at least six hours to infuse the plant into the water. You can strain the liquid and/or add a sweetener such as honey to any type of tea. Sample teas include:

- Common cold: Echinacea
- Energy/alertness: Ginseng and ginkgo
- High blood pressure: Hibiscus (to lower it)
- Improve immunity: Rosehip
- Nausea: Ginger
- Relaxation: Chamomile, lavender, or rosemary
- Sore throat: Licorice root
- Upset stomach: Peppermint

Tinctures

Tinctures have a much longer shelf life than teas (months to years) and can incorporate a wider variety of plants. They work best with dried plants, but you can use fresh ones too.

To make a tincture, first sterilize a glass jar with boiling water.

Pulverize the plant as much as you can in any way you want, then pack it into the glass jar. Cover the plant matter with 80% proof alcohol. Vodka is common, but other alcohols, like whiskey or rum, will also work also. Different types will give the tincture different flavors. Ensure the plant matter is fully submerged.

Seal and store the jar in a cool, dark place for six weeks. Shake it up every couple of days.

After six weeks, strain out the plant matter through cheesecloth and discard it. Put the medicinal liquid in a tinted glass container and store it in a cool, dark place. Label it.

The standard adult dose for many tinctures is one teaspoon up to three times a day.

Sample tinctures include:

- Chamomile (flower): Anxiety, healing wounds, and reducing inflammation
- Echinacea: Boosting immune function
- Feverfew (leaf): Reducing fevers, preventing migraines and treating arthritis
- Ginger (root): Treating nausea, and motion sickness
- Gingko (leaf): Treating asthma and tinnitus, improving memory, and preventing dementia
- Peppermint: Aiding digestion
- St. John's wort (flower, leaf): Treating depression
- Valerian (root): Treating insomnia

FIRE CIDER TINCTURE

This makes a tincture using apple cider vinegar (ACV) to extract the medicinal value. It has antibiotic, antiviral, anti-fungal, and many other qualities.

What You Need

- Apple cider vinegar.
- Hot chili peppers. Remove the stems.
- Garlic.
- White onion.
- Ginger root.
- Turmeric root.
- Horseradish root.
- Lemon. Do not peel it.
- Blender.
- Raw honey.
- 2 glass jars, sterilized with boiling water.

As a general guideline, use the dry ingredients in roughly equal parts, or do it to taste. For example, if you don't want it super hot, use less chili. After making it a few times, you'll know what you like.

Use the freshest ingredients you can find. If you're missing a few ingredients, it's okay. Use what you have. Only the apple cider vinegar is integral.

Directions

Wash, peel, and roughly chop all the dry ingredients, then blend them all together.

Pack the blended ingredients into a sterilized glass jar and cover them with apple cider vinegar.

Seal the jar and put it somewhere cool and dry. Shake it daily for a minimum of two weeks. The longer you leave it before straining, the stronger it will be.

Strain the liquid through cheesecloth into the second glass jar. Stir in the raw honey to taste and store it in a cool, dark place.

Shake the jar well before using the tincture. Drink one tablespoon mixed with filtered water once a day for preventative purposes. Have the same dosage up to every three hours when you are sick.

PEST CONTROL

Pests such as cockroaches, mosquitoes, and rodents spread disease. Here are some things you can do to prevent them coming into your living areas.

Keep your home clean. If there is nothing left out for them to eat and drink, they will (hopefully) stay away.

Stuff entry points (pipes, small holes, etc.) with steel wool.

Get a pet. Cats are better at keeping pests away than dogs.

Plant insect-repelling plants, though they only work as far as their odor carries. You can make a barrier of plants by planting them less than 1m (3ft) apart. Some types you can use are:

- Flies: Lavender, basil, rosemary
- Moths: Lavender
- Mosquitoes: Basil, lavender, mint, rosemary, lemon grass, marigolds
- Termites: Hot chili peppers, mint, vetiver grass

COCKROACH AND ANT KILLER

This is a natural paste that attracts and kills cockroaches and ants. They will also take it back to their colony so it has a widespread effect.

The active ingredient in the paste is boric acid, which is a derived from the natural element boron. Boric acid powder is dangerous to inhale, so when you are making this, make sure you wear a mask and are in a well-ventilated area. This paste is harmful to most living things. If you have pets and/or small children, make sure they cannot get to the bait.

What You Need

- Boric acid powder. If you can't find pure boric acid, get a product with as high a percentage as possible.
- Any sugary syrup, like honey, maple, or sugar water.
- Flour. Any type will work.
- A bowl.
- A popsicle stick.

Directions

Mix two parts boric acid powder with one part flour and enough sugary syrup to make a thick paste. Use the popsicle stick to apply it in places where cockroaches like to go, such as:

- At the back of drawers
- In crawl spaces
- In pipes coming into your home
- In the attic
- Inside outlet covers
- Near water
- Up high

Apply it to ant entry points as well.

The paste won't stick until it's dry, so put it on level surfaces where it can sit.

If the ants won't stop, you may need to head outside and kill the colony. Pour hot, soapy water into their nest.

MOSQUITO OVITRAP

This mosquito trap is cheap and easy to make. It uses no poison and kills off the next generation of mosquitoes too.

What You Need

- A black plastic container.
- A black sock.
- Silicone-based glue. It must be waterproof.
- A metal mesh screen. You want the mesh sized so sand can pass through, but a mosquito can't.
- Stagnant water. Make stagnant water by putting some organic material, like glass clippings, compost, or dry dog food, in a bucket of water.
- Scissors.
- Twine or wire (optional).

Directions

Create a hole in the container about a fifth of the way down from the top. This is the drain hole. Glue the toe end of the sock to bottom of the container.

Cut a circle out of the mesh screen, using the top of the container as a template.

Once the glue is dry, pull the sock over the mouth of the container. Inside the container, the sock needs to be taut, but not super tight. Glue it in place and wait for it to dry.

Insert the mesh screen into the top of the container. It must sit above the drain hole.

Fill the container with stagnant water. The water must be below the bottom of the metal screen. The drainage hole will take care of this.

Place it where the mosquitoes like to hang out, like under garden lights, but not too close to where you like to hang out. You do not want to attract the mosquitoes to you.

If you want to hang it, make a couple of holes opposite each other at the top and use some wire or twine to make a handle.

The mosquitos are attracted to the stagnant water, and they lay their eggs on the moist dark sock. As the eggs hatch, the larvae go into water below. They hatch into live mosquitoes and are too big to get through the screen.

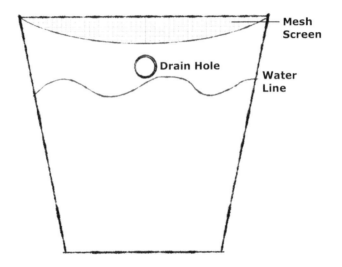

You need at least five ovitraps per acre to start bringing the mosquito population down. It will be a couple of weeks before you start to see real results. Keep them filled with stagnant water.

When a trap gets full, pull out the screen and empty it. Put the screen back and refill with it with stagnant water to catch more mosquitoes.

RODENT BUCKET TRAP

It is not good to have rodents such as mice and rats running around, but many of the options to get rid of them are cruel. This simple trap will catch a small rodent without hurting it. Then you can decide to either release it away from your home or kill it humanely.

Be careful when handling rodents as they may carry diseases. Use rubber gloves and wash thoroughly afterwards. If you are bitten or scratched, seek medical attention ASAP. Don't keep any mouse or rat you catch as a pet.

What You Need

- A bucket with a lid. It must be big enough that the rodent won't be able to climb out.
- Wire. Use a coat hanger or the wire handle from the bucket.
- A sharp serrated knife.
- A hammer and a nail.
- Sticky rodent bait, such as peanut butter.
- A stick.
- Dirt.

Directions

Cut the flat part of the lid out with the serrated knife. Keep it as an intact circle and as large as possible. This is the trapdoor.

If you have a bucket with no lid (or you don't want to destroy the lid), you can use a paper plate or something similar instead as long as it is the right size.

Use the hammer and nail to punch a hole in the trapdoor big enough for the wire to fit through. Make the hole close to the edge, then do it on the opposite side too.

Make two holes at the top of the bucket in the same way. Thread the wire through the two holes in the trapdoor and bucket so the door is "skewered" and fixed to the bucket.

Put a layer of dirt in the bottom of the bucket. This helps weigh the bucket down and softens the rodent's landing.

Put some bait on both sides of the trap door in the center, then lean the stick against the bucket for the rodent to climb.

The rodent will climb up the stick to get the bait. When it steps on the trapdoor, it will fall into the bucket.

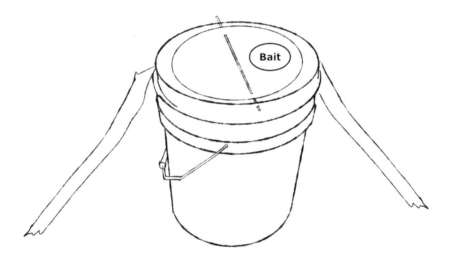

Release the rodent at least 5km (3mi) from your home, or it may find its way back.

To adapt this for larger pests, such as raccoons, use a bigger bucket and the right bait.

THANKS FOR READING

Dear reader,

Thank you for reading *DIY Sustainable Home Projects*.

If you enjoyed this book, please leave a review where you bought it. It helps more than most people think.

Don't forget your FREE book chapters!

You will also be among the first to know of FREE review copies, discount offers, bonus content, and more.

Go to:

https://offers.SFNonfictionBooks.com/Free-Chapters

Thanks again for your support.

REFERENCES

Bourn, C. (2016). *Living Off The Grid: The Essential Guide to Embracing Minimalism and Self Reliance with Your Own Sustainable Homestead.* Chase Bourn.

Chillingsworth, J. (2019). *Live Green: 52 Steps for a More Sustainable Life.* Quadrille Publishing.

Faires, N. (2016). *The Ultimate Guide to Natural Farming and Sustainable Living: Permaculture for Beginners.* Skyhorse.

Liu, C. (2018). *Sustainable Home: Practical projects, tips and advice for maintaining a more eco-friendly household.* White Lion Publishing.

Schifman, M. (2018). *Building a Sustainable Home: Practical Green Design Choices for Your Health, Wealth, and Soul.* Skyhorse.

Toht, D. (2013). *40 Projects for Building Your Backyard Homestead: A Hands-on, Step-by-Step Sustainable-Living Guide (Creative Homeowner) Fences, Chicken Coops, Sheds, Gardening, and More for Becoming Self-Sufficient.* Creative Homeowner.

AUTHOR RECOMMENDATIONS

Teach Yourself Evasive Wilderness Survival

Discover all the evasive survival skills you need, because if you can survive under these circumstances, you can survive anything

Get it now.

www.SFNonfictionBooks.com/Evasive-Wilderness-Survival-Techniques

This is the Only Wilderness Medicine Book You Need

Discover what you need to heal yourself, because a little knowledge goes a long way

Get it now.

www.SFNonfictionBooks.com/Wilderness-Travel-Medicine

ABOUT SAM FURY

Sam Fury has had a passion for survival, evasion, resistance, and escape (SERE) training since he was a young boy growing up in Australia.

This led him to years of training and career experience in related subjects, including martial arts, military training, survival skills, outdoor sports, and sustainable living.

These days, Sam spends his time refining existing skills, gaining new skills, and sharing what he learns via the Survival Fitness Plan website.

www.SurvivalFitnessPlan.com

- amazon.com/author/samfury
- goodreads.com/SamFury
- facebook.com/AuthorSamFury
- instagram.com/AuthorSamFury
- youtube.com/SurvivalFitnessPlan

CPSIA information can be obtained
at www.ICGtesting.com
Printed in the USA
LVHW080401290622
722373LV00017B/291